PROGRAM MANAGEMENT FOR IMPROVED BUSINESS RESULTS

PROGRAM MANAGEMENT FOR IMPROVED BUSINESS RESULTS

Second Edition

Russ J. Martinelli
James M. Waddell
Tim J. Rahschulte

Published by John Wiley & Sons, Inc., Hoboken, New Jersey
Published simultaneously in Canada

For general information about our other products and services, please contact our
Customer Care Department within the United States at (800) 762-2974, outside the
United States at (317) 572-3993 or fax (317) 572-4002.

Wiley publishes in a variety of print and electronic formats and by print-on-demand.
Some material included with standard print versions of this book may not be included
in e-books or in print-on-demand. If this book refers to media such as a CD or DVD
that is not included in the version you purchased, you may download this material
at http://booksupport.wiley.com. For more information about Wiley products, visit
www.wiley.com.

Library of Congress Cataloging-in-Publication Data

Martinelli, Russ J., 1959-
    Program management for improved business results / Russ J. Martinelli,
James M. Waddell, Tim J. Rahschulte. – Second edition.
        1 online resource.
    Revised edition of Program management for improved business results by
Dragan Z. Milosevic, Russ J. Martinelli, James M. Waddell, published in 2007.
    Includes index.
    Description based on print version record and CIP data provided by publisher;
resource not viewed.
    ISBN 978-1-118-90587-6 (ePub); ISBN 978-1-118-90589-0 (Adobe PDF);
ISBN 978-1-118-62792-1 (hardback); 978-1-118-90436-7 (O-book) 1. Project management.
2. Project management–Case studies. I. Waddell, James M., 1946- II. Rahschulte, Tim.
III. Milosevic, Dragan Program management for improved business result. IV. Title.
    HD69.P75
    658.4'04–dc23

                                                                    2014019692

Printed in the United States of America

V10010280_052119

To our dear friend Dragan Milosevic

Contents

Part V: Organizational Considerations   303

Appendices: Case Studies in Program Management   349

Preface

Since the publication of the first edition of this book, many volumes of white papers, articles, and books on the subject of program management have emerged. The impact, as measured by increased knowledge about *what* program management is and *why* it is important, has been great. We feel fortunate to have been part of that change.

We also feel fortunate to have received some amazing feedback from the readers of the first edition, which was both complementary and constructive. The most rewarding feedback came from readers who felt the book helped them to become better program managers (or become first-time program managers in some cases), as well as from those who recognized that the book is "different". This book *is* different by design. The differentiator is that its foundation is based upon a *body of practice* that focuses on *how* program management has been practiced historically within companies, and *how* it is practiced today.

Our personal understanding of how program management is being practiced has been greatly enhanced by opportunities to train many practitioners in a variety of industry sectors (both for-profit and non-profit), and by opportunities to work directly with companies that are engaged in the introduction of program management into their organizations or that are working to strengthen their existing program management capabilities. This new understanding, and the associated lessons learned, are shared throughout this second edition.

The most significant changes introduced in this edition are in four areas. First, we introduce the concept of the program management continuum, which we use as an anchor throughout the book to describe the variation of how program management is implemented within companies, and how we delineate between project-oriented and program-oriented organizations. Next, we provide a broader explanation of the relationship between systems thinking and program management, to include one of the primary roles of a program manager as the master integrator of cross-project work. Then, to strengthen one of the emerging themes of program management, we include additional information on benefits management, particularly in relation to the achievement of the business benefits that drive investment in programs. Finally, we worked to create tighter alignment and cross-reference to the program management standards and guides that have been developed to provide additional detail and depth to the program management principles.

To reinforce the practical nature of this book, we include seven new case studies. Four case studies, referred to as Program Management in Practice, are found at the end of each major section of the book, and three comprehensive case studies that focus on multiple dimensions of program management are included in the appendices. The case studies represent the application of program management in a variety of industries, including software services, automotive, academia, information technology, U.S. defense, and digital media display. We chose to use fictitious names for the companies and people presented in the case studies to ensure the good, the bad, and the ugly aspects of each case remained intact. The cases are real, however, as are the characters and the stories contained within. In our choice of tools to present, we cover those that we see utilized the most and those that provide the greatest utility. Additional tools and tool templates can be found on the Program Management Academy website: http://wiley.programmanagement-academy.com.

Finally, we maintained the modular design and flow of information contained in the first edition. This allows you, our readers, the option to read this book from cover to cover, or to focus upon the aspects of program management that are most pertinent to your needs. However you choose to read this book, we hope you enjoy your journey into the world of program management.

Acknowledgments

We would like to thank the many people who have helped in making this book a reality:

David Churchill, Richard Vander Meer, Roger Lundberg, Rick Nardizzi, Gary Rosen, Manny Mora, and Richard Cook for their invaluable viewpoints, opinions, and insights about program management as it is practiced in companies today.

Ronald Forward for his significant contributions in sharing his knowledge, experience, and insights in regard to leading and managing a Program Management Office in a major corporation.

Kathy Milhauser and Eddie Williams for contributing their experiences in the form of case studies.

Steve Graykowski for inspiring the idea of the program management continuum and thought leadership involved with integrating program management and agile software practices.

David Pells, who continues to provide the means to test our ideas with the readers of the PM World Journal.

Margaret Cummins, Amanda Shettleton, and the remainder of the team at John Wiley & Sons who continue to provide outstanding support and guidance.

Our many colleagues and co-workers who have contributed to the concepts presented in this work in many ways.

We are truly blessed to be associated with such a wonderful and supportive community of people!

Part I

It's About the Business

Part I begins by providing clarification of the program management discipline and then illustrating how program management can be implemented as a major part of an organization's business model.

The primary theme established in this first part, and then used throughout the entire book is *it's about the business*. The purpose of the introductory chapter, Program Management, is to establish the foundational elements of programs and program management as it is practiced in our organizations and many of our clients' organizations, and explain how it is used to achieve a firm's strategic business goals. The unique meaning of program management is identified and described, illuminating its raison d'être. It explains what program management is and what it is not and compares and contrasts program management with project management and portfolio management, dimension by dimension.

The foundational elements from Chapter 1 provide perspective for Chapter 2, Realizing Business Benefits. In our own careers, we have witnessed the power of program management to serve as a coalescing function that provides business benefits by delivering both business *value* and business *results*. In Chapter 2, we explore these two sides of business benefits realization through the implementation of program management within an enterprise.

Chapter 3, Aligning Programs with Business Strategy, completes Part I by detailing the systematic approach of program management through the use of an integrated management system. As we demonstrate in this chapter, the program management discipline plays a pivotal role in aligning the work output of multiple project teams to the corporate and business unit strategy of an enterprise.

Chapter 1

Program Management

A lot has changed with respect to program management since we introduced the first edition of this book. Much of the literature that existed at that time consistently confused program management with project, portfolio, or operations management. Today, multiple standards exist and many volumes of white papers, articles, and books are readily available. As a result, the general knowledge about *what* program management is and *why* it is valuable has increased markedly.

While many different aspects and approaches to program management have emerged, we have been pleased to watch a convergence on what we believe is the single most important aspect of program management: it's about achieving business results.

Even the various standards, which by nature take a broad brushstroke at the subject of program management, state that program management is all about benefits realization, and benefits directly refer to achievement of the business goals of the enterprise and the organizations within the enterprise.

The purpose of this introductory chapter is to establish the foundational elements of programs and program management *as it is practiced in our organizations and many of our clients' organizations,* and explain how it is used to achieve a firm's strategic business goals.

This is the foundational information needed by anyone considering the introduction of program management within their organization, or for anyone needing a better understanding of how their current use of program management can be further matured to gain improved business results and establish a stronger link between execution and strategy.

DEFINITIONS AND CONTEXT

One of the primary challenges with creating standards, and therefore standard definitions, is that they have to be broad in nature to encompass a wide range of applications, but specific enough that individuals can identify and correlate the work they do on a day-to-day basis within the standard.

Programs Defined

The two leading standards with respect to program management are of course *The Standard for Program Management* by the Project Management Institute (PMI) in the U.S., and *Managing Successful Programmes* by the Office of Government Commerce (OGC) in the U.K. From an academic and standards perspective, each of these organizations has created a useful but differing definition for a program.

PMI defines a program as "a group of related projects managed in a coordinated manner to obtain benefits and control not available from managing them individually. Programs may include elements of related work outside of the scope of the discrete projects in the program."[1]

The OGC defines a program as "a temporary flexible organization created to coordinate, direct and oversee the implementation of a set of related projects and activities in order to deliver outcomes and benefits related to the organization's strategic objectives."[2]

From a practice standpoint, we can utilize either definition, depending on the organization we are working with and their particular view of what a program encompasses. There are a few points of particular interest, however, that we tend to point out regarding these definitions when teaching or coaching. We particularly like the fact that the PMI standard identifies that a program includes "elements of related work outside of the scope of the discrete projects." As we explain in Chapter 4, by taking a *whole solution* or systemic approach to defining and structuring a program, one quickly realizes that a program needs to encompass more than the constituent projects within the program to be truly successful.

Additionally, the OGC standard brings out the fact that programs exist "to deliver outcomes and benefits related to the organization's strategic objectives." This is a critical distinction in practice: programs must exist to further the strategic business goals of an enterprise. Otherwise it becomes work for the sake of doing work—a result that unfortunately is all too common in many organizations.

Finally, each standard describes a program as consisting of a group or a set of "related" projects. We would rather the standards be a bit more precise regarding this point. If the projects are merely related, what distinguishes a program from a portfolio in these definitions? In practice, the projects within a program have a higher level of relationship. They are not just related, but rather highly *interrelated*. The distinction here is that each project is so dependent upon one or more of the other projects on the program that it cannot succeed on its own. If one of the projects on a program fails, it is highly likely that the program in its entirety will fail. This is an important distinction because it is not necessarily the case for a portfolio of related projects.

Understanding these subtleties with regard to the definitions for a program will help in the application of the term within your organizations.

Program Management Defined

While we were writing the first edition of this book, a common, universally accepted definition of program management did not exist. When we researched the definition we found many versions that were similar in some ways and quite different in other ways. Interestingly, the same is true today, only there are fewer versions available.

Although we have slightly refined our original definition of program management, we continue to find that it is most effective for people who are either implementing program management into their organizations or looking to mature their existing program management culture and practices.

Program Management Definition

Achieving a set of business goals through the coordinated management of interdependent projects over a finite period of time.

This definition describes a model of program management that exists within an organization that has a high degree of program management maturity, what we call a *program-oriented organization*. Within a program-oriented organization, program management exists as a critical element within the business operations of the enterprise. It is in this context that the maximum gain will be realized from the existence and practice of program management. For this reason, this definition contains a number of key tenets, each of which are addressed below.

Benefits Management

Benefits realization through the achievement of an organization's business goals is the overriding objective of any program, and therefore the management of a program. For this reason our definition of program management begins with this realization: "achieving a set of business goals." By way of example, in product or service development, a key program-level goal is to introduce capabilities before one's competitors. In a competitive environment, time-to-benefits is arguably the most closely tracked metric by both the program manager and senior management. We do not dispute that delivery of the right product at the right time is critical, especially since we have had plenty of personal experiences where that was the primary measure of success. However, delivery of the product is only the mechanism to realize the *true business goals*, such as capturing additional market share, increasing profit through sales and gross margin growth, and strengthening brand value by being the first to market with compelling features and usages.

Coordinated Management

Most programs require the work of many functions within an organization. Therefore they must be organized into a set of project teams that are cross-discipline and cross-functional. Using the phrase "coordinated management" of multiple projects in our definition means that the activities and outcomes of each project team are executed through a common program framework and synchronized by the program manager. Steven Wheelwright and Kim Clark properly articulated the need for effective cross-functional management many years ago:

> Outstanding development requires effective action from all of the major functions in the business. From engineering one needs good design; from marketing, thoughtful product positioning, solid customer analysis, and well-thought-out product plans; and from manufacturing, capable processes. But there is more than this. Great products and processes are achieved when all of these functional activities fit well together. They not only match in consistency, but they reinforce one another.[3]

Interdependent Projects

For program management, cross-discipline and cross-functional coordination and integration has to be extended to include *cross-project* coordination and integration. Every program is made up of multiple projects, each of which is most likely cross-discipline in nature. This concept is described by Mary Willner, a senior manager at Intel Corporation:

> With one set of desired business results for the program, coordination extends beyond just schedule coordination; it also requires coordination to ensure the stated business objectives are met. Which, if compromises are required (e.g. cost, feature, schedule), its resolution is managed as a coordinated effort across the interdependent projects.[4]

As the term implies, "interdependent projects" are those that have a mutual dependence on the output of other projects in order to achieve success. Commonly, the interdependencies come in the form of deliverables that are the tangible outputs from one project team that become the input to another project team or teams. Program management ensures that the dependencies between the multiple projects are managed in a concerted manner.

Finite Period of Time

A "finite period of time" means that a program is a temporary undertaking, having a point of beginning and a point of ending. This can be contentious as some definitions describe a program as an ongoing endeavor. From our perspective, if this is the case the program is really part of the normal *operations* of the business, therefore not a discrete program and may be better defined as an initiative. By contrast, for a program in which a new capability or organizational change is created and delivered, the program must have both a beginning and an ending in order to effectively measure business results.

This point came to the forefront when we were asked to assist a leading customer relationship management software company with the implementation of program management into their product development and IT businesses. The company historically has had an agile development and delivery culture, which caused much debate to ensue among the senior leaders of the company on whether a program should have an end, or rather should be a continuous process in the spirit of the agile methodology. The debate ended with the realization that a program did in fact have to end in order to measure whether the business goals driving the need for a program were achieved. In this case, the programs that deliver new capabilities to their software platform or into their IT infrastructure are time-bound (usually a year or six months).

PROGRAM MANAGEMENT CHARACTERISTICS

A definition alone does not provide adequate description of the value that program management can bring to an enterprise. There are five core tenets underlying program management practices that help to describe the true value of program management as a unique business function.

Establishes Ownership and Accountability

In many organizations that do not utilize the program management model, ownership and accountability for the business results associated with the program normally falls on the functional managers of the business. Generally, in project-oriented organizations ownership and accountability of a program can pass from one functional group to another—for example, from research during the concept phase, to marketing during the feasibility phase, to engineering during the planning and execution phases, to manufacturing during the production readiness phase, and finally, back to marketing for capability release. Passing the ownership baton can work well in a perfectly conceived, planned, and executed project, but quickly breaks down when problems begin to surface and personal accountability is required on the part of one or more of the functional managers. With a program management model, there is no debate or subjectivity about who owns, and is accountable for, the business success or failure of the program; the program manager assumes this full responsibility throughout the program cycle.

Strategic in Nature

The program management discipline helps to ensure that a program is closely aligned to, and directly supports, the achievement of a business's strategic goals (Chapter 3).[5] In effect, it is used to direct the activities involved with the *implementation* of strategy (see "Turning Strategy into Action at Intel"). Figure 1.1 illustrates the link between program management and business strategy.

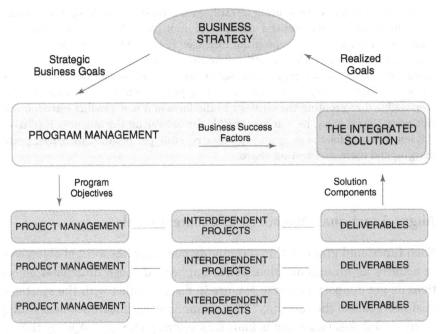

Figure 1.1 The strategic nature of program management.

Program management links execution to strategy by integrating the work flow and deliverables of multiple interdependent projects to develop and deliver an integrated solution. This integrated solution becomes the means by which the strategic goals of the business are achieved.

Turning Strategy into Action at Intel

An example from a well-known technology company provides a good illustration of how program management is used to deliver strategic business results.

Several years ago, Intel Corporation developed a strategic goal to increase the amount of revenue it received from each of its microprocessor products by providing added capabilities within the product. One of the strategies set in motion to achieve this goal was to adopt a platform approach that would serve to integrate various technologies into a single product.

One of the original platforms developed integrated both computing capabilities and wireless communications technologies into a single integrated circuit. Legacy solutions involved a microprocessor to handle computing and a separate component, or add-in card, for wireless communication for a personal laptop computer. Intel's *strategy* to achieve this technology convergence resulted in the development of a new family of microprocessors that combine the two technologies. The market now knows the resulting product as Centrino™.

Critical to effectively executing its strategies is Intel's use of the program management discipline to direct the activities of multiple technology-specific project teams. In this example, the program manager responsible for the development and launch of the new Centrino™ microprocessor was responsible for the integration of the computing project, wireless project, multiple software projects, the test and validation project, the manufacturing project, and more. In short, he was responsible for executing the strategy in the form of a new product introduction. But more importantly, he was responsible for delivering the business objectives driving the need for the program—greater revenue per product, increased profit margin, and increased market share.

Aligns Functional Objectives to Business Goals

Each functional group within a company normally has a set of objectives to achieve as an organization. What happens if these functional objectives do not support, or worse yet, are in direct conflict with the strategic business goals of the company? This dilemma is a difficult problem facing many businesses today and is known as **agency theory**.[6] Agency theory occurs when functional managers design objectives that provide the greatest benefit for their own organization and consider the strategic goals of the company as a secondary consideration.

Program management can be used to reduce the negative effects of agency theory by aligning functional objectives to corporate and business unit objectives. In a program-oriented enterprise, the various business functions take on a different role than in either a functional- or project-oriented organization. In a program-oriented organization, the functions exist to support the achievement of business objectives, which are realized through the successful execution of the organization's programs. Therefore, much of the *functional group's* success is dependent upon *program* success. This change in how success is measured serves to effectively align the business functions not to functional success, but rather to program and business success.

Fosters Cross-Project and Multi-Disciplined Integration

Programs, by design, are cross-project in nature in that they involve multiple projects that are coherently and collectively managed to achieve the program output. Additionally, the constituent projects of a program are normally centered on individual disciplines within an organization, such as design, engineering, customer support, or marketing. To reconcile the cross-project, multi-disciplined nature of programs, many organizations employ a matrix structure to span the various functions needed to effectively implement a program. Program management becomes the common thread that sews the matrix together and enables the cross-project teams to perform cohesively. Organizationally, program management provides the opportunity to manage work efforts across the traditional (hierarchical) line structure of an organization, thus contributing to faster decision-making and improved productivity.

Enables Distributed Collaboration

A new business model has emerged where knowledge work is digitized, disaggregated, distributed across the globe, produced, and reassembled again at its source.[7] Team collaboration can now occur without regard to geographical boundaries or distances. Companies that are thriving in this new business model are the ones that are successfully integrating this distribution of work. Because of its relationship to systems thinking, program management has emerged as an effective method for managing such work. Within a program, the program manager synchronizes the work of the project teams as they create their respective pieces of the whole solution, and then works to integrate the output of the specialized knowledge workers into a total, efficient solution.[8]

The backdrop of defining program management now coupled with these five tenants—ownership and accountability, strategic focus, alignment of objectives, cross-project and multidiscipline, and distributed collaboration—characterizes the business value of program management to the organization.

THE PROGRAM MANAGEMENT CONTINUUM

In reality, not all companies utilize their program management discipline as an extension of their business processes. Therefore, the role of the program manager can vary greatly from company to company (or even within different divisions of the same company).

In organizations where the level of program management use is high, it is often viewed as part of the business management function and is linked to corporate and business unit strategy. We refer to these organizations

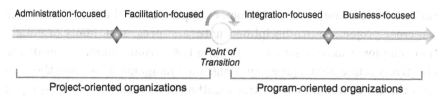

Figure 1.2 The program management continuum.

as being *program-oriented*. Conversely, in organizations where program management is not used to its full capacity, program management normally serves as a coordination function and an extension of project execution. We refer to these organizations as *project-oriented*.

Figure 1.2 illustrates program management orientation as a continuum with four distinct stages: administration-focused, facilitation-focused, integration-focused, and business-focused.

In general, organizations that are primarily *project-oriented* utilize program management practices as either an administrative or facilitative function. As one moves to the right of the center point on the continuum, we find *program-oriented* organizations that realize the potential of program management to serve as an integrator of project outcomes that is at least linked to the business engine of the company.

Details on the various stages of the program management continuum reveal the varying view and use of program management, and therefore the variation of roles and responsibilities of program managers.

An ***administration-focused organization*** demonstrates a strong focus on single projects, and very strong functional or line management control of the projects. The program management function is quite limited within these organizations and is utilized primarily as an administrative, data-gathering, and activity-monitoring function.

A ***facilitation-focused organization*** is also project-oriented, but the projects are normally grouped into programs, usually organically rather than strategically. Program management serves as a coordination function that facilitates cross-project communication and low levels of collaboration. It is typical to find organizations using portfolio management techniques when operating at this point on the continuum, with programs often serving as subportfolios instead of true programs.

An *integration-focused organization* views projects as part of a program which is driven from business strategy. At this point on the continuum, control of the projects shifts from the functional or line managers to the program managers. The primary focus of program management is the integration and synchronization of work flow, outcomes, and deliverables of multiple projects to create an integrated solution.

A *business-focused organization* is fully devoted and disciplined in its use of program practices. Programs are tightly linked to strategy and serve as the execution mechanism to realize business goals. In the business-focused culture, organizational hierarchical command and control is replaced by empowerment and accountability on the part of the program manager.

Organizations typically evolve the use of program management practices and as such, move from left to right along the continuum as the needs of the business demand. However, as an organization moves from left to right (increasing its program management maturity), it is not relieved of the program management responsibilities it has left in the earlier stages of the continuum. Rather, the responsibilities are cumulative. Even if one reaches the final stage of maturity and is operating as a business-focused program management organization, there are still a number of administration, facilitation, and integration duties to fulfill.

At the center of the continuum lies an important point we refer to as the program management *point of transition*. The point of transition is a philosophical decision point where the senior leaders of an organization make a purposeful and concerted choice to move their organization from being primarily project-oriented to being program-oriented. The transition point represents the formal acknowledgment by senior management regarding the importance of the program management discipline as a strategic benefit to the organization and the need to formally and actively empower the program manager to fulfill this role.

We do not advocate that all organizations should move across the point of transition and become program-oriented. There are many organizations that are quite successful and operate effectively as project-oriented enterprises due to the nature of their business. They should continue to do so and look for aspects of program management that are beneficial to the way they operate.

Frog, a leading product strategy and design company, has been on a journey of increasing program management maturity for a number of years. Richard Vander Meer, Vice President of Global Program Management, described the company's journey:

> When we created the program manager role, we hired people to be note-takers and meeting schedulers. But as the needs of the company changed, we needed to create a more robust function. Today, our program managers are the main interface between the client and the program team, and they have P&L (profit and loss) responsibility for the client engagement program.

For the companies that realize the need for transition from one level of program management maturity to another, they must do so knowing that crossing the point of transition creates considerable change in culture and requires a clear vision and strong leadership. In general, the following changes may be needed:

- Roles and responsibilities realignment
- Skills and competency development
- Talent acquisition
- Team and organization structure adjustments
- Strategy adjustments
- Shifts in decision empowerment
- New practices and success measures
- Different incentives and individual performance indicators

We cover the organizational impacts of crossing the *point of transition* in detail in Chapters 12 and 13. The intent of discussing them here is to give you a broad understanding of the changes that need to be in place organizationally if your goal is to realize the full power of program management.

The program management continuum is used throughout this book to demonstrate the variations in the implementation of program management within companies today. The concepts in this book are centered on the practices of program-oriented enterprises. However, there is one important point we want to make before moving on to another topic. It is critical for the senior leaders, functional and line managers, program managers, and project managers within a company to realize where they are on the program management continuum in order to align expectations, and properly set roles and responsibilities as well as empowerment and decision-making boundaries for the various actors. As the example titled "An 'Ah-ha' Moment" illustrates, it is common that we find a difference of opinion where an organization sits on the continuum.

An "Ah-ha" Moment

The senior management team of a major U.S. defense contractor was situated in one of the company's training rooms and engaged in a lively discussion about the performance of their program managers. At the center of the discussion was a question posed to them about the program manager's role in driving business results for their organization.

Eventually the discussion shifted to the individual abilities of each of their program managers to fulfill an expectation that they operate as the CEO of the program they are managing. At that point Frank DeYoung, R&D Director for one of the product line businesses, reset the discussion by stating, "This really isn't about the abilities of each of our program managers. I think it's about setting clear expectations and then performing to those expectations."

With that, the discussion moved to the Program Management Continuum and a detailed description of organizational culture, roles, responsibilities, decision making, and empowerment for each of the stages of the continuum. The senior leaders were asked where on the continuum they *expected* their program managers to be operating. Lively discussion again ensued, with consensus suggesting the leadership team expected the program managers to be operating in the far right stage of the continuum, "As a business manager," it was mentioned.

They were then asked another question: Where do you believe the program managers currently see themselves operating? This time the discussion was more subdued because they realized a problem was about to surface. The leaders agreed that the program managers were currently operating to the left of the center point of the continuum (see Figure 1.3). The gap between expectation and reality was now apparent.

A long period of silence was finally interrupted by DeYoung, who stated, "I see the problem, and this team owns fixing it." DeYoung later described this as his "Ah-ha Moment" about program management.

So, why is there a gap between expectations of how the program managers should operate within this company and how they are actually performing? The

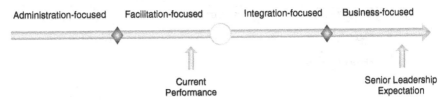

Figure 1.3   **The gap between expectation and reality.**

senior leadership team characterized the gap as having two primary causes. The first cause was described as an atrophy of their program-oriented culture, which was due to prior disinvestment of their program management office and much of the infrastructure that was in place to sustain and enable the program management discipline.

The second cause was the senior leadership team themselves. Each of the leaders had been a program manager on his or her journey to senior business leader roles. However, they had those roles at an earlier time when the organization was fully program-oriented and they were in fact able to operate as true business managers in charge of a program. The realization was that they expected their program managers to operate in like manner, but the organization's culture no longer supported it.

This company is not unique. Nearly every organization we work with has a gap between where they are currently operating with respect to program management and where they want to operate going forward. This is the program management journey to improved benefits realization, a journey that is often set in motion to realize improved business results and over time requires resetting to assure expectations are aligned and being met.

THE RELATIONSHIP AMONG PORTFOLIO, PROGRAM, AND PROJECT MANAGEMENT

For program-oriented organizations, the relationship between portfolio, program, and project management is normally well defined and understood. As illustrated in Figure 1.4, portfolio management is a method used to prioritize programs according to the overall business strategies of the

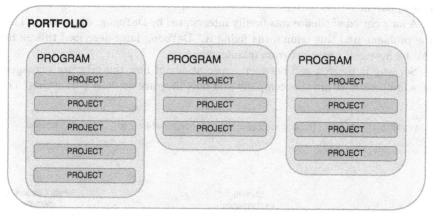

Figure 1.4　Portfolio, program, and project management relationship.

organization, while program management is responsible for the actual execution and delivery of those strategies.

Further, the delivery of the strategies is accomplished through integration of a multitude of outcomes that are a result of the management of multiple projects within a program. The relationship between portfolio, program, and project management is summarized as follows:

- A *portfolio* consists of multiple programs that represent an organization's investment in the achievement of its strategic goals. The programs within the portfolio are the mechanisms used to execute the strategic goals defined at the enterprise or business unit level.
- A *program* consists of a set of interrelated projects whose outcomes are integrated at the program level to create a whole solution that may take the form of a product, service, or change state that becomes the means to achieving business goals.
- A *project* is a component of a program from which a set of work outcomes or deliverables contribute to creating the whole solution. It is at the project level that the tangible goods are created by discipline-specific work teams. Additionally, projects also exist as stand-alone work efforts that are independent of programs.

Program-oriented organizations have a strong relationship between portfolio, program, and project management. This relationship is not only in place, it is often well defined and understood across the organization. These organizations have the opportunity to consistently ensure that the tangible goods generated at the project level are effectively integrated at the program level and the resulting solution is in alignment with the strategic goals of the business.

In contrast to program-oriented companies, project-oriented companies often have a limited relationship between portfolio management, program management, and project management. Because of this, it is common for confusion to exist between the three forms of management. The next two sections delineate the major differences between program management and project and portfolio management.

Differentiating Program and Project Management

Table 1.1 provides a summary of the important differentiating factors between program and project management. The primary differentiator is the fact that program management is *strategic in nature and focused on*

Table 1.1 Program and project management differentiation.

Differentiating Factor	Program Management	Project Management
Strategic versus Tactical	*Strategic* in nature, focused on business success	*Tactical* in nature, focused on execution success
Alignment	Aligned to the business goals of the organization	Aligned to the program objectives
Responsibility	Successful delivery of the entire integrated solution	Successful delivery of a portion of the integrated solution
Work Effort	Assures the cross-project work effort remains feasible from a business standpoint	Assures work effort generates deliverables on time, within budget and at required performance levels within the project's specialty
Risk	Concerned with cross-project risk affecting the probability of program and business success	Concerned with single-project risk affecting the probability of project success
Life Cycle	Involved in all stages of the program cycle, from definition to end of life	Primarily involved in the planning and implementation stages of the program cycle
Process Orientation	Ensures consistent use of common processes by all project teams	Ensures effective and efficient implementation of processes on a single project
Control	Monitors and controls the delivery of cross-project deliverables	Manages and controls the tasks associated with development of project deliverables
Change	Navigates change as it is encountered and resets the program to changes in business goals	Controls change to an established project baseline

the delivery of an integrated solution, while project management is *tactical in nature and focused on the successful execution of a portion of the integrated solution.* All other factors in the summary are subfactors of this primary differentiator.

Alignment

Managing a program means ensuring that the program remains in alignment with, and in support of, the strategic goals set forth by senior management. This includes alignment with the organization's strategic plan, its portfolio and roadmap, and the business-related goals such as financials, market penetration, and technology advancement. The project manager, in turn, is responsible for ensuring the work and resulting deliverables of their project are in alignment with and in support of the program objectives.

We refer to project management as tactical in nature based on PMI's Project Management Body of Knowledge (*PMBOK®*) as the dominant industry practices standard where project management is about management of a single, individual project, whose primary focus is accomplishment of the triple-constraints (time-cost-scope).[9]

Responsibility

On a program, the program manager's job involves the successful delivery of an integrated solution that requires management of the interdependencies across the multitude of projects. By way of example, if an engineering project team encounters a quality issue that will impact the timing of their deliverable to the manufacturing project team, the program manager must determine if it is better to delay the deliverable (and the work of the manufacturing project team) or reduce the quality target. This is a cross-project issue to be solved at the program level. In contrast, the project manager is focused on the scope of work within a single project and is responsible for the successful delivery of the outcomes of the project to the program.

Work Effort

A program manager work effort focuses heavily on the integration and synchronization of the work outcomes of the constituent projects on the program. In contrast, the work effort of the project manager focuses only on project-centric deliverables.

Risk

Similar to the scenario above, both the program manager and the project managers are responsible for identifying and managing risk on a program, but do so in different dimensions. Program risk management involves identifying and managing cross-project risks that may affect the overall probability of business success of the program.[10] Project risk management, on the other hand, involves identifying and managing risks that may affect the probability of technical success for a single project (see Figure 7.3 in Chapter 7 for an illustration of this concept).

Life Cycle

Life cycle in this context pertains to all of the stages that a program progresses through from the time of its inception to its eventual closure. In a program-oriented organization, program management is involved in all stages of the life cycle. This includes the definition, planning, execution, operational, and closure stages. By contrast, project management is typically associated with the planning and execution work cycles.

Process Orientation

From a process perspective, the distinction between program and project management is in how processes and procedures are established and executed. The program manager is responsible for ensuring that company processes and procedures are established on the program, and that they are consistently used by all project teams. The project manager is responsible for effective and efficient implementation of the processes and procedures established by the program manager, as well as those established by the managers of functional organizations for their particular discipline.

Change

Critical to project management is the establishment of a baseline from which to effectively execute. Any change introduced is normally tightly controlled with a penchant for change avoidance in order to prevent rework and drive assurance of the scope and timeline. Critical to program management, however, is awareness of change occurring in the business environment that will affect the success of the program. Program managers must be adept at navigating change and understanding the impact of change on the business goals driving a program.

Control

As with independent projects, project management on a program involves monitoring and controlling the progress of the tasks being performed to create the project deliverables. Program management is focused on a level higher, which involves monitoring and controlling the synchronization of deliverables between the project teams on a program in support of creating an integrated solution.

DIFFERENTIATING PROGRAM AND PORTFOLIO MANAGEMENT

At times, confusion also exists between program management and portfolio management. One of the causes of this confusion may be that they are often both broadly defined as the management of multiple projects. This is, however, where the similarity ends. This section provides a brief characterization of portfolio management for readers who are not familiar with the process and describes the key distinctions between portfolio management and program management.

Characterizing Portfolio Management

The senior management team of an organization utilizes the portfolio management process to synthesize current and future collective intelligence of the organization to select, prioritize, fund, and resource the portfolio of opportunities that will best achieve the attainment of the strategic goals. In synthesizing the intelligence of the organization, various key factors about the business and business environment must be analyzed to obtain the right mix and number of opportunities. Such factors may include the following:

- Company strategic objectives
- Customer wants, needs, and usage requirements
- Competitive intelligence
- Current and future technology capability of the enterprise
- Risks and potential rewards
- Resources and other assets available to plan and implement the portfolio[11]

The portfolio management process is an ongoing process that ensures a company is working on the opportunities that offer the highest probability for attractive financial and strategic returns at the lowest possible risk. Opportunities are ranked and prioritized based upon a set of criteria that represent *value* to the organization. Resources are then allocated to the highest value and most strategically significant opportunities. Low-value opportunities must be cut, returned for redefinition, or put on hold until adequate resources become available.

Summary of Program and Portfolio Management Differentiation

Table 1.2 provides a summary of the important differentiating factors between program management and portfolio management. The primary

Table 1.2   Program and portfolio management differentiation.

Differentiating Factor	Program Management	Portfolio Management
Process versus Function	A management *function* utilized to determine the business and execution feasibility of a single idea, and then turn the idea into an actionable plan that is successfully executed and delivered to the customer	A *process* utilized to evaluate, prioritize, select and resource a collection of new ideas that best contribute to the attainment of the strategic goals of an organization
Value	Focused on ensuring the business value is attained for a single opportunity within a portfolio	Focused on determination of the business value of all existing opportunities of the organization
Risk	Concerned with cross-project risk affecting the probability of program and business success	Concerned with balancing risk and return for the aggregate portfolio of opportunities
Resources	Ensuring a single opportunity is adequately staffed with the right resources (number, skills, experience)	Aligning an organization's resources to opportunities that provide the greatest strategic value to a business

differentiator is the fact that *portfolio management is a decision-making process, while program management is a key management function within an organization*. All other factors in the summary are subfactors of this primary differentiator.

Value

The heart of the portfolio management process is the ability of the senior management team to determine the business value of the various opportunities available to the company. Therefore, the portfolio management process identifies the critical factors that determine opportunity value (common factors were noted previously in this section).[12]

Once the business value is determined for an opportunity within the portfolio, and the opportunity is selected for funding and resource allocation by the senior management team, the opportunity is assigned to the program management function within the enterprise. Program managers are then responsible for turning each of the portfolio ideas into a tangible outcome and *delivering* the value to the senior management team.

Risk

The senior management team manages portfolio risk from both a macro and micro perspective. Macro-level risk management of a portfolio involves determining the overall risk level of the aggregate opportunities within the portfolio, then determining the right balance of opportunities based upon the risk tolerance of the organization. From a micro-level perspective, senior management, along with the appropriate knowledgeable members of the organization, must assess each key element of the portfolio in order to balance the portfolio risk against the potential reward.

Once an opportunity is funded, the program management function assumes ownership of the management of the risk for the duration of the program. As stated earlier, management of risk at the program level is focused on overall probability of achieving the business goals driving the need for the program.

Resources

Businesses typically have more ideas than human and non-human resources to carry them out. As a result, resources can become overcommitted and weighed down by an overwhelming list of opportunities to pursue. Portfolio resource management involves aligning resource demand to capacity, and assigning resources to opportunities that provide the greatest value to a business. The end result of a well-executed portfolio management process is a balance between high-value opportunities and the number of available resources to execute those opportunities.

Upon approval and funding of an opportunity, efficient and effective resource management for the development of the opportunity becomes the responsibility of the program manager and the functional managers of the organization. In order for the value of an opportunity to be realized, the program designed to deliver the opportunity must be adequately staffed with the correct number of resources that possess the right skills required and the appropriate level of experience.

IS PROGRAM MANAGEMENT A NEW CONCEPT?

Interestingly, many view program management as a relatively new phenomenon. We surmise this is a matter of perspective. If one's perspective is related to the work of the various project management organizations such as PMI, OGC, and IPMA (International Project Management Association), your perspective certainly may be that program management came to light in the first decade of the twenty-first century. However, if one has been involved in practicing program management for a number of years, the perspective may be different. Program management in fact has formally existed in companies for over six decades. The first documented evidence of program management in the U.S. dates back to the 1950s (see "On Origins of Program Management").

Regardless of its history, the program management discipline has now fully emerged from its early practices and is being broadly adopted across both for-profit and nonprofit industry sectors. Various authors have attributed this to the current business environment, which can be described as dynamically changing, ambiguous in nature, and more complex than ever before. They recognize that traditional management practices have limitations with this new business environment and that program management practices are well suited to provide the necessary means to integrate project outcomes with business strategy in fluid situations with high levels of complexity.

On Origins of Program Management

The exact origin of program management is not definitively known. The U.S. Military argues that they were the first to have developed and implemented program management. The Manhattan Project was the first to use program practices (in the 1940s) to create the atomic bomb.[13] Program management practices were also said to be used (in the 1950s) on the Atlas Program to create the first intercontinental ballistic missile.[14] Our own research shows the first documented evidence of program management dates back to 1957 with the formation of the first program office, then called the Special Project Office (SPO), within the United States Department of the Navy.[15] The SPO was established to manage the development of an underwater ballistic missile launch system. Indeed, the structure of the missile launch system program mirrors the program management structures utilized today—a series of interrelated projects (launcher, missile, guidance, installation, navigation, operations, and test) collectively and coherently managed as a program. In the early 1970s, the program management discipline became popular across the United States Department of Defense and the SPO became the first program management office.

On July 1, 1971, the doors of the Defense Management School, later called the Defense Systems Management College (DSMC), opened at Wright-Patterson Air Force Base to admit the first students enrolled in the 20-week program management course.[16] The original mission of the DSMC was to: 1) conduct advanced courses in study of program management; and 2) assemble and disseminate information concerning program management. In 1993, the name was again changed, to the Defense Acquisition University (DAU) to reflect a new mission and broader scope of academic study and research in program management.[17] Today, thousands of military and military support personnel graduate from DAU annually.

Until the 1980s, the program management discipline and the DSMC that resided within the military and defense industries were well-kept secrets. During this time period, companies that maintained both defense and commercial businesses, such as Boeing, Lockheed, and other aerospace companies, began migrating the program management discipline and management model from their military divisions to their commercial divisions. Program management proved to be very effective in the management of complex product development efforts. Today, the program management discipline and practices continues to expand throughout many commercial and private industries.

ENDNOTES

1. Project Management Institute. *A Guide to Program Management Body of Knowledge*. Newtown Square, Penn.: Project Management Institute, 2004.

2. Office of Government Commerce. *Managing Successful Programmes.* 3d ed. Norwich, UK: Office of Government Commerce, 2007.

3. Wheelwright, S., and K. Clark. *Revolutionizing Product Development.* New York: Free Press Publishing, 1992.

4. Willner, M. Personal interview, 2007.

5. Martinelli, R., and J. Waddell. "Demystifying Program Management: Linking Business Strategy to Product Development." *PDMA Visions* magazine, 2004.

6. Pearce, J., II, and R. Robinson, Jr. *Strategic Management: Formulation, Implementation, and Control.* New York: McGraw-Hill Publishing, 2010.

7. Friedman, Thomas L. *The World Is Flat.* New York: Farrar, Straus and Giroux Publishing, 2006.

8. Martinelli, R., T. Rahschulte, and J. Waddell. *Leading Global Project Teams: The New Leadership* Challenge. Oshawa, Ontario: Multi-Media Publishing, 2010.

9. Project Management Institute. *A Guide to Program Management Body of Knowledge.* Newtown Square, Penn.: Project Management Institute, 2004.

10. Martinelli, R. and Jim Waddell. "Managing Program Risk." *Project Management World Today,* September–October 2004.

11. Cooper, R. G., S. J. Edget, and E. J. Kleinschmidt. *Portfolio Management for New Products.* Cambridge, Mass.: Perseus Publishing, 2001.

12. Ibid.

13. Weaver, P. The Origins of Modern Project Management, 2007. Retrieved January 19, 2013 from www.mosaicprojects.com.au/PDF_Papers/P050_Origins_of_Modern_PM.pdf.

14. Ibid. Additional information about Atlas can be found at http://en.wikipedia.org/wiki/SM-65_Atlas.

15. Ashie, Ibrahim A. Department of the Navy Strategic Systems Programs Office, www.dau.mil/pubs/pm/pmpdf94/ashie.pdf.

16. Defense Acquisition University Press. *U.S. Department of Defense Extension to: A Guide to the Project Management Body of Knowledge,* June 2003, Version 1.0.

17. Summers, Wilson. "Before DSMC, There Was DWSMC." *Program Manager,* January–February 2000.

Chapter 2

Realizing Business Benefits

It's about the business. That is the primary theme running throughout this book, and for program-oriented organizations, the realization of business benefits is the primary purpose of their program management discipline. In our own careers, we have witnessed the power of program management to serve as a coalescing function that focuses the various elements of an organization upon the achievement of its business goals and is the primary reason why we titled this book *Program Management for Improved Business Results*.

Many others seem to agree. The achievement of business benefits is the center of the program management guides and standards that have emerged and matured over the years, and has been a primary topic of writings from other authors. From our viewpoint, however, there needs to be additional distinction on what is now commonly referred to as business benefits in regard to program management. The business benefits that program management helps to deliver comprise two components: business *value* and business *results*. Business value encompasses the synergistic improvements that program management brings to optimize the business functions of an enterprise. Business results are the tangible business outcomes that result from the creation and delivery of new capabilities through the firm's programs.

In this chapter, we explore these two sides of business benefits realization through the implementation of program management within an enterprise.

REALIZING BUSINESS VALUE

Benefits realization in the broad sense is closely tied to *value management*. Ray Venkataraman and Jeffrey Pinto described value management as "a management approach that focuses on motivating people, developing skills, and fostering synergies and innovation with the ultimate goal of optimizing the overall organizational performance."[1] In a sense, they are describing the value that program management brings to an organization, especially to those organizations that implement program management to its full capacity. In this case, business value is accomplished through the optimization of a firm's business functions in a number of ways, including:

- Aligning business strategy and execution
- Integrating business functions
- Navigating business and environmental ambiguity
- Achieving business scalability
- Managing distributed collaboration
- Reducing time-to-benefits

Aligning Business Strategy and Execution

Many organizations engage in yearly strategic planning activities that focus on identifying long-range business goals, as well as high-level plans on how to achieve them. Good strategic management practices identify *what* an organization wants to achieve (strategic goals) and *how* they will be achieved (strategies) over a specified time horizon, which is typically three to five years. For product companies, strategy consists primarily of a collection of product ideas that, when turned into tangible products, contribute to the achievement of the strategic business goals. For service-oriented companies, strategy consists of a set of services that collectively contribute to the achievement of the strategic goals, and for change initiatives, strategy consists of new organizational transformation and breakthrough capabilities that will help achieve company strategic goals.

As an organization begins to grow and scale its offerings, maintaining alignment between work output and strategic intent often becomes a challenge. In part this is due to the increased number of offerings in the pipeline, but also the increasing number of cross-organizational groups required to be involved, and in many cases, an increasing layer of middle management. Simply stated, the link between execution output and strategic goals begins to weaken, and in some firms may eventually break.

For many organizations, program management provides value to their business by serving as the organizational *glue* that can prevent the misalignment between execution and strategy. As detailed in the next chapter, program management creates a critical linkage across strategic goals, program objectives, and project deliverables focused on delivering the business benefits intended. In doing so, one can establish direct traceability between strategic goals, business success factors, and project performance measures.

Integrating Business Functions

Implementation of strategy involves many business functions and processes that need to be coordinated and integrated into synchronized business actions. Traditional project- or functional-oriented approaches possess some of this integration capability. However, they tend to be tactically focused on the triple constraints—time, cost, and scope—and many times fail to serve as the business integrator focused on the implementation of strategy.

Integration efforts require significant collaboration across multiple organizations and functions, which can be hampered by parochial behaviors within traditionally siloed functions. Compounding this problem is that many times, the management personnel involved in strategic planning, portfolio planning, and execution are different with limited overlap, which minimizes effective communication and collaboration. This results in ineffective business integration.

Program management adds business value by serving as the mechanism by which the work of the various operating functions within a company is integrated to create an effective business model. For example, consider the business functions of marketing, engineering, manufacturing, and finance. Each function has its own language and jargon. Marketing language talks about the four Ps (product, price, place, promotion), finance discusses discounted cash flow, engineering discusses technical performance, and manufacturing is interested in production yield and defect rates. To say that experts from different functions often do not understand one another, let alone often do not work well together, is an understatement.

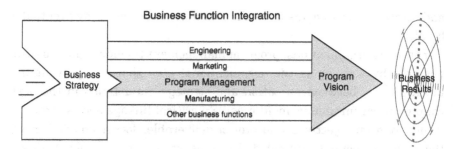

Figure 2.1   Integrating business functions.

The inclusion of program management in a firm's business model provides value to those organizations seeking to integrate the efforts of their business functions to achieve business results through the development and delivery of new capabilities or transformative change. When program management is properly defined and implemented, it helps to execute business strategies through the collective efforts of the business functions of an enterprise. Figure 2.1 illustrates this point.

Program management integrates the collective efforts by focusing the various functions on a common purpose: achievement of improved business results. Even though they may speak in differing jargon, commonality is established through a collective purpose and goal. The integration of cross-organizational efforts increases the probability of successfully achieving the intended business results in a *repeatable* manner.

Navigating Business and Environmental Ambiguity

Management teams in successful and innovative companies fully understand that some of the greatest opportunities reside in the *fuzzy front end* of a program.[2] The ability to accurately forecast future customer, user, and market needs and then integrate those needs with leading edge capabilities is critical for companies to survive in their respective industries.[3] This work is never simple and high levels of ambiguity have served as a test in frustration and a lesson in patience for many.[4]

However, the presence of ambiguity is a characteristic of many capability development efforts, and the ability to manage it effectively can provide great business opportunity and competitive gain. Program management can provide business value for companies looking to successfully produce competitive capabilities on a consistent basis by providing the means to effectively navigate ambiguity and harness the innovation engines of their firms.

The key to realizing these opportunities is the ability to contain the fluid and ambiguous nature of a program. This can be accomplished by establishing a framework and well-defined targets to provide focus and by employing effective leadership to cut through the ambiguity and competing agendas of stakeholders that often characterize a program.[5]

Richard Vander Meer, Vice President of Global Program Management, described the role of program management in containing ambiguity at Frog:

> I see program management as the backbone of the company. We have to maintain stability in the middle of a very creative and often chaotic design-driven environment. This involves establishing and maintaining integrity of work, ensuring financial responsibility, and providing stability for the team and the client.

Containing ambiguity through a framework involves creating a system of objectives that focus and motivate the efforts of the program team. The objectives must be both time- and business-focused. Time-focused targets consist of regularly scheduled review meetings with senior management to demonstrate continued convergence progress in the development of a capability concept. Business-focused targets include the relevant objectives that drive the need for the capability (for example, increased revenue, lower operating costs, or market transformation).

By employing program management principles to contain ambiguity, an organization will realize the following three key benefits: 1) the leadership necessary to effectively coalesce multiple perspectives and agendas; 2) a framework to enable flexible management of change; and 3) a business champion to ensure that the strategic goals set forth by senior management are achieved.

Achieving Business Scalability

As an organization grows, its ability to effectively scale its business processes and consistently deliver positive return on investment becomes increasingly challenging. The challenge normally falls upon the business manager charged with investing in new capabilities. When an organization is small, the business manager is able to manage both the investment in new capabilities as well as manage the return of the investment

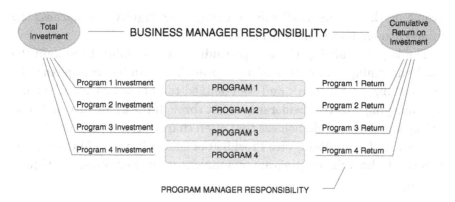

Figure 2.2   **Achieving scalability with program management.**

through the creation of new capabilities. When higher levels of growth occur, however, the business manager often finds that he or she is not able to perform both management functions as effectively as needed.

Program management offers business value by providing the organization a pathway to effectively scale the business by assuming the management of return on investment duties for the development of each new capability.

As illustrated in Figure 2.2, a business manager's ability to scale a business becomes constrained by his or her ability to personally scale. The addition of the program management discipline and skilled program managers removes this constraint and allows for business scalability.

The business manager still maintains total investment responsibilities for his or her business and for each program, as well as accountability for *total* return on investment. However, the business manager allocates portions of his or her investment to the various programs within the portfolio and to the program managers responsible for managing the investment through execution of a program. The program managers become accountable for delivering the return on investment (business results) for each of their respective programs. In effect the program manager serves as the *business manager proxy*.

A prerequisite for business scalability through program management, however, is the level of business acumen of the program managers. To achieve scalability, the business manager must be able to hand over the management of return on investment with confidence. Therefore, the program manager must possess the skills and competencies to fully manage the business aspects of a program (Chapter 11).

Managing Distributed Collaboration

Due to the amount of complexity required to meet customer and user demands for performance, features, and customized capabilities, today's work efforts are often beyond the scope of a single project.[6]

Effectiveness requires an *integration* of work efforts and outcomes to satisfy the growing complexity of business. However, another modern phenomenon—distributed teams—has added an additional layer of complexity and difficulty to the integration of multiple work outcomes.

The world is getting flatter. This is how Thomas L. Friedman, in his book titled *The World Is Flat*, describes the phenomenon that began in the 1990s and continues today, whereby knowledge work can be digitized, disaggregated, distributed, produced, and reassembled across the globe.[7] The flattening of the world has enabled people in countries like China, Russia, India, and many others to participate in capability development efforts in the west. This has created a new business model where highly distributed collaboration is required.

However, many companies have historically operated under a traditional structure characterized by strong siloed departments or groups where horizontal collaboration across these departments is difficult, let alone collaboration across the globe. One by one these companies are realizing the need to adopt a distributed model not only to compete, but in many cases to survive.

Many companies that are succeeding in management in these increasingly distributed environments are doing two key things: 1) adopting a systems approach to developing their capabilities, and 2) adopting program management to effectively integrate their solutions. Early adopter companies in the automotive, aerospace, and defense industries continue to utilize this approach. More recently, companies such as Apple, salesforce.com, Intel, and Kaizer Permanente among others have found great success in utilizing systems and platform concepts coupled with program management to develop their new capabilities. Not surprisingly, these companies are consistently succeeding in the management of distributed efforts that employ outsourcing, open-sourcing, in-sourcing, and off-shoring techniques.

Reducing Time-to-Benefits

Besides demanding increasingly complex solutions, consumers also want accelerated delivery of new capabilities. It is a well-known fact that

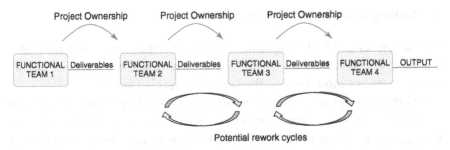

Figure 2.3 Project hand-off approach.

in today's highly competitive world, time-to-benefits is a critical factor in gaining an advantage. For most companies, gaining time-to-benefit advantage means decreasing the cycle time required for delivering a new capability. Historically, the two most dominant approaches were the project hand-off and concurrent development methods. While each of these approaches has merit in simple synchronous efforts, most organizations today are challenged when using these approaches due to their inherent limitations and constraints relative to complexity.

In the *project hand-off approach*, each functional team sequentially works on their element of the project, then hands both the work output and project ownership over to the next functional team in line. This approach is illustrated in Figure 2.3.

The limitations of such work are explained in the example entitled "The Perils of the Project Hand-off Approach," the primary problem of which is that the work accomplished at each hand-off occurs within a single function. Errors introduced upstream can only be reconciled downstream, resulting in multiple rework cycles that consume time-to-benefit advantage.

The Perils of the Project Hand-off Approach

Hospi-Tek is a medical equipment manufacturing company that has historically used a project hand-off approach to develop its products. They are currently under intense time-to-market pressure from their primary competitor, forcing senior management to reevaluate their approach.

Under the project hand-off method, the Hospi-Tek product development effort began with the architectural team, which developed an architectural concept and derived the high-level requirements of the medical device from the work of the product marketing team. The architectural concept and specifications were then handed off to the engineering team, which assumed ownership of the project.

The engineering team developed the hardware requirements, engineering specifications, and the product design, which were then handed off to the manufacturing team, which assumed ownership of the project. The manufacturing team developed the manufacturing processes, retooled the factory, and produced the physical product. The product and project ownership were then handed off to downstream engineering teams, including the software development team. The software team developed the software stack, then handed the combined hardware/software product, as well as project ownership, to the validation and test team. Finally, the validation and test team performed product and component-level testing to ensure the product achieved the functional, quality, usability, and reliability requirements.

Management of the project was accomplished through a project management-only model, with multiple project managers in control of the project as it progressed through the development life cycle. Thus, a project manager with the functional expertise specific to the phase of development the product was currently in assumed ownership of the project.

This method of development is common in smaller and technically focused companies in which true project and program management value is usually not well understood and the engineering function reigns king. Unfortunately, this method is not scalable, and as a company begins to succeed and grow, product and process complexity requires the management team to look at alternative methods to structure and manage its development efforts. This was the case with Hospi-Tek.

The *concurrent development approach* originated to decrease cycle time over the project hand-off approach.[8] It involves the various functional teams working simultaneously to deliver their elements of the capability under development (Figure 2.4). The concurrent development approach,

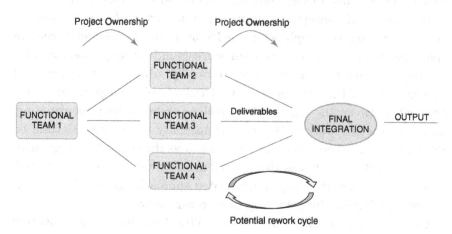

Figure 2.4   Concurrent development approach.

however, has been found to be less than optimal in reducing development cycle time due to the *big bang* event that inevitably has to occur in concurrent development.

The big bang occurs when the concurrently developed work is integrated. Great inefficiencies can occur due to rework caused by poor requirements definition, lack of change management in the functional development efforts, and poor communication between the concurrent development teams. As a result, development teams may spend considerable time performing integration, testing, and rework due to these inefficiencies caused in an earlier cycle. Depending on the extent of any misalignment of the functional outputs being integrated, the concurrent method may actually lead to an *increase* in cycle time over a project hand-off approach.

Having now outlined two historic approaches and their limitations, a logical question is, "How do companies achieve a significant reduction in cycle time?" The best-in-class companies that we have observed, researched, or have worked for accomplish reduced cycle times through an *integrated* approach.

In the integrated approach, the cross-discipline teams work shoulder to shoulder during the work effort, from conception through end of life. To increase the probability of time-to-benefit efficiencies, the historically back-end functional activities such as manufacturing, validation, and testing are pulled forward in the work flow. Additionally, the front-end activities such as design and marketing are involved longer to ensure seamless integration. The *big bang* phenomenon is replaced by continuous, iterative-design, develop, and integrate cycles (Figure 2.5).

In order for the integrated approach to achieve its full potential, the cross-project, multi-discipline team is led by a program manager who can span the disciplines to ensure their work is indeed fully integrated for each cycle of work. Achieving time-to-benefit advantage requires tight management of the interfaces, shared risks, and open communication channels between the teams. This involves orchestrating, coordinating, and directing the work of the various specialty teams (see "Getting on the Fast Track"). The program manager brings the general management, business, and leadership skills required to effectively lead the integrated, cross-project team. By using an integrated approach, an organization has a higher probability for achieving time-to-benefit goals.

One company leader, Gary Rosen, Vice President of Engineering for Applied Materials Corporation, agreed with this assessment and

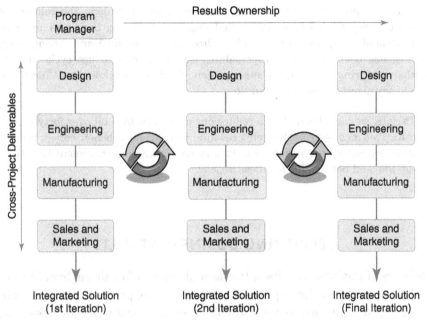

Figure 2.5 The integrated approach.

described the role of program management in accelerating time-to-benefits:

> If you have a strong program management function, products are closer to what the customers want and the team spends less time iterating late in the program to meet customer expectations. A program manager adds clarity for the engineering team by balancing market requests with engineering capabilities, therefore setting realistic customer targets. This results in more efficient use of resources which allows a program team to deliver what the customer wants the first time.

Getting on the Fast Track

Following the advice of a well-known consulting firm, Hospi-Tek applied a concurrent approach to develop their next two medical equipment products in an effort to reduce cycle time. Much of the previously sequential project work was pulled forward and performed in parallel. However, Hospi-Tek struggled when significant misalignment between project elements (hardware, software, and manufacturing) became visible during the integration stage of the project. They concluded that the integration process takes nearly as long as the up-front design and development processes.

With a 5 percent reduction in cycle time achieved through the concurrent approach, they moved to an integrated approach, utilizing their new program management discipline. The Hospi-Tek Director of Program Management, Terry Levins, has been given the responsibility to continually reduce cycle time by further maturing the program management discipline in their firm. Terry had this to say about the current situation:

> To date, we have been able to reduce our cycle time by a bit over forty percent. This means we begin seeing revenue flow (business benefit) about nine months sooner than in the past. This is a significant improvement to both top line and bottom line business benefit.

DELIVERING BUSINESS RESULTS

Delivery of business results is the second aspect of business benefit realization attained by the implementation and use of program management. To reiterate, business results are the tangible business outcomes that result from the creation and delivery of new capabilities through the use of program management.

PMI identifies a benefit as an outcome of actions and behaviors that provides utility to an organization,[9] while the OGC sees benefits as the tangible business improvements that support a firm's strategic objectives.[10] Both organizations therefore recognize that programs and projects deliver benefits by enhancing current capabilities or developing new capabilities that support the sponsoring organization's strategic goals. A benefit is therefore a contribution to a strategic goal of a firm and in many cases cannot be fully realized until after the completion of the program. An important distinction being made here is that new or improved capabilities as the output of a program are the means to an end, and the anticipated business results define the end state.

There is a direct correlation, however, between how program management is implemented within an organization and the amount of business benefit realized through the discipline.

We discussed in Chapter 1 that companies using program management apply it over a range of implementations based upon their company culture and program management philosophy and maturity.

Figure 2.6 illustrates that for organizations that are more project-oriented, the potential for business results being driven from the program

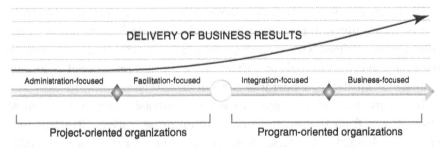

Figure 2.6   **Delivering increased business results.**

management discipline normally is relatively low. Rather, the business results are often the responsibility of the business unit manager or of a department manager.

As a business makes a conscious decision to become more program-oriented, the responsibility for delivering business results begins to fall upon the program manager.

This has much to do with the different types of programs found within project- and program-oriented organizations and the difference in how they are created.

Generally, project-oriented organizations utilize formulated programs that are formed from the combination of preexisting projects and other work activities into a single entity. In this program type, a realization that the projects may be more effectively managed under a single program occurs. This realization is many times driven by a desire on the part of a firm's executive team to take a more strategic approach to the work that gets accomplished within their organization.

Critical to the formulated program type is the establishment of a common purpose or goal that ties the various project and work activities together. If this does not exist, it should be recognized that the organization most likely has a portfolio of projects rather than a true program composed of interdependent projects and the projects therefore should continue to be managed independently or consolidated in some other way.

Formulated programs require a clear vision be established early on to precipitate the identification of which projects should fall within the program, thus defining the scope of the program. Along with a clear program vision, direct linkage of the program to the appropriate strategic goals of the business is required. This enables the program team to map their project outcomes and deliverables to the desired business results.

In the case of formulated programs, it is common that investment in the various individual projects is made before the decision to invest in the program as a whole. This causes investment reconciliation to occur once the program is formulated since these types of programs are not originally spawned from strategic business goals of an enterprise.

In program-oriented organizations, by contrast, programs are most often driven from the strategic goals of an organization. Strategic programs are defined by strategic objectives, and therefore it is more straightforward to define the business results desired from the creation and delivery of the program output. The qualifier to this statement is that the level of ease is dependent upon how well the firm's senior managers define their strategic goals.

In most cases, an investment is made in a strategic program before the projects are fully scoped and planned. Based on this distinction, strategic programs are not defined from the bottom up by the scope of their constituent projects, as some definitions of a program suggest. Rather, they are defined by a top-down view of the whole solution that will enable achievement of the strategic goals.

Benefits Delivery and the Program Manager

Program-oriented organizations view the program manager as the business unit manager's proxy for managing the return on investment by delivering the business results anticipated. Some even go so far as viewing the program manager as the CEO (Chief Executive Officer) of a specific program. Few would argue that one of the main responsibilities of any CEO is the realization of improved business results.

To be effective, the program manager will need to keep the focus of a program on the business results, not solely on the delivery of a new or improved capability. This is an important distinction. Often a program manager gets so focused on the creation and delivery of a new product, service, or change transformation capability that they lose sight of the fact that these are the means to the end as discussed previously. This is not to say that the creation and delivery of a capability is not important. In fact, it is *critically* important. Without the capability there is no means to achieving improved business results. Our point, however, is that the capability in itself is not the primary goal. The goal is the business result that the new capability enables.

ENDNOTES

1. Venkataraman, Ray, R., and Jeffrey K. Pinto. *Cost and Value Management in Projects*. Hoboken, N.J.: John Wiley & Sons, 2008.

2. Smith, Preston G., and Donald G. Rinertsen. *Developing Products in Half the Time: New Rules, New Tools*, 2d ed. Hoboken, N.J.: John Wiley & Sons, 1998.

3. Koen, P. A., G. M. Ajamian, et al. "Fuzzy Front end: Effective Methods, Tools, and Techniques." In *The PDMA ToolBook for New Product Development*, P. Belliveau, A. Griffin and S. Somermeyer. New York: John Wiley & Sons, 2002, pp. 5–35.

4. Cooper, Robert G. *Winning at New Products: Accelerating the Process from Idea to Launch*, 3d ed. Cambridge, Mass.: Perseus Books, 2001.

5. Martinelli, Russ. "Taming the Fuzzy Front End." *Project Management World Today* (July–August 2003).

6. Martinelli, R., Tim Rahschulte, and James Waddell. *Leading Global Project Teams: The New Leadership Challenge*. Oshawa, Ontario, Canada: Multi-Media Publishing, 2010.

7. Friedman, Thomas L. *The World Is Flat*. New York: Farrar, Straus and Giroux Publishing, 2006, pp. 439–440.

8. Belliveau, P., A. Griffin, and S. Somermeyer. *The PDMA ToolBook for New Product Development*. New York: John Wiley & Sons, 2002, 5–35.

9. Project Management Institute. *The Standard for Program Management*, 2d ed. Newtown Square, Penn.: Project Management Institute Publisher, 2008.

10. Office of Government Commerce. *Managing Successful Programmes*, 3d ed. Norwich, UK: Office of Government Commerce Publisher, 2007.

Chapter 3

Aligning Programs with Business Strategy

Historically, the primary managerial functions and processes of a company have been defined and viewed as independent entities, each with its own purpose and set of activities. For example, executive management normally performs the strategic processes that set the course of action for the organization. Portfolio management and project selection are commonly thought of as senior and middle management responsibilities. Program planning and execution processes are performed by the program manager and program team, while project managers and team leaders are responsible for project planning and execution processes. Each of these functions and processes are executed separately by a different set of people within the organization. At best, the strategic element feeds the portfolio element, the portfolio element feeds the program management element, and the program management element feeds the projects and specialty team execution. In many cases, this results in projects that may not be tied directly to either the business strategy or the organization's portfolio due to the lack of process integration.

Companies have come to realize that the time, money, and human effort invested in refining and improving each of their independent functions and processes have not brought them closer to effectively and efficiently turning their ideas into positive business results. Increasingly, this fact is leading business leaders to the realization that their independent entities can no longer remain independent if they wish to repeatedly achieve their desired business benefits and business value. Rather, they must be transformed into a set of *interdependent* elements that form a coherent strategy and business benefit realization system.

By taking a systematic approach through an integrated development model, an organization can realize improved business alignment between execution output and business strategy. As we demonstrate in this chapter, the program management discipline plays a pivotal role in the alignment process.

THE INTEGRATED MANAGEMENT SYSTEM

When one looks at an entire business from a systems perspective, a holistic view emerges. Within the enterprise system, key subsystems such as the corporate mission, strategic goals, organizational functions, organizational structure, critical processes, and programs and projects exist to effectively and efficiently convert the resource inputs into the desired strategic outputs, such as corporate growth, increased productivity, or effective organizational transformation. Like any system, the subsystems are highly interdependent upon one another. For example, the mission of the business enterprise influences the strategic business goals defined, and the strategic goals define the business functions needed and how they should be organized and interact.

Additionally, the business enterprise operates within, and is influenced by, a dynamic environment. Examples of environmental factors that have an impact on the mission, structure, operation, and output of the business enterprise system include shareholder expectations, domestic and world economic conditions, technology trends, customer usage models, and competitor actions.

The heart of the enterprise is a system of management functions and critical processes needed to convert inputs into outputs. We refer to this set of functions and critical processes as the *integrated management system*.[1] Illustrated in Figure 3.1, this system is the mechanism from which new products, services, infrastructures, or organizational transformation initiatives are conceived and developed to realize the mission, strategic goals, and business benefits of the enterprise.[2]

It is important to view the integrated management system as a coherent end-to-end system from understanding market and environmental conditions, to the development of strategic goals that support a company mission, to the delivery of business value. This end-to-end structure allows for more flexibility and receptivity to changes in the dynamic business environment.

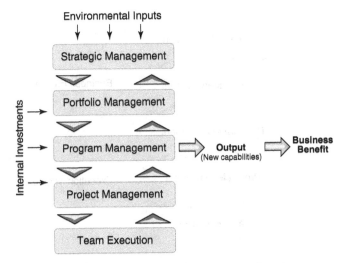

Figure 3.1   The integrated management system.

By contrast, taking a noncoherent approach is analogous to actions one may take if an automobile is performing poorly and exhibiting below-par performance symptoms, such as low fuel efficiency or rough idling. By taking an ad hoc approach to the problem, one may try to treat the low fuel efficiency by adding a fuel oxidization additive or replacing the fuel filter. To treat the poor idling problem, one may change the spark plugs or adjust the engine timing. Any or all of these actions may yield an improved performance for a period of time, but they will not solve the root problem if the automobile is in need of an overall tune-up. Only by viewing the engine and ignition functions of the automobile as a system does a holistic approach to diagnosing and resolving the root problems become possible.

With this analogy in mind, we view the integrated management system as being comprised of two primary components: 1) the business engine and 2) the execution engine (Figure 3.2). The business engine consists of the strategic management, portfolio management, and program management element of the business. The execution engine consists of the team execution, project management, and again the program management elements. Program management is the overlap between the business and execution engines of the business and serves as the organizational glue that aligns strategy and action, which is necessary to deliver business value.

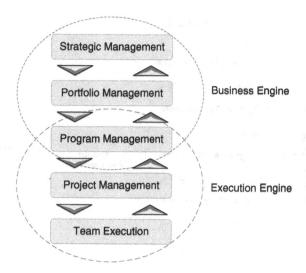

Figure 3.2   Business and execution engines of an organization.

THE BUSINESS ENGINE

The business engine of the integrated management system is composed of three primary functions: strategic management, portfolio management, and program management. Strategic management involves definition of the company mission, identification of strategic goals, and creation of strategies to achieve the goals. Portfolio management includes the review and selection of the strategic options (programs) to be implemented, evaluation of the success of the strategic process in achieving the organizational objectives, and alignment of the organization's funding and resources to the most strategically important programs. The program management element of the business engine involves translating strategic goals into solution possibilities and execution requirements, developing and validating a business case, and delivering the business benefits through creation of new capabilities.

Strategic Management

Strategic management is defined as a set of decisions and actions that result in the formation and implementation of plans designed to achieve a company's objectives.[3] The mission statement is a broadly framed statement of intent that describes *why* the company exists in terms of its purpose, philosophy, and goals. Strategic goals define *what* the business

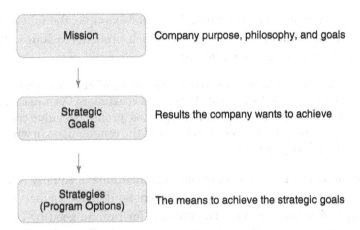

Figure 3.3   The strategic management process flow.

wants to achieve over the strategic time horizon, which is typically three to five years. Strategy is the business' game plan that reflects *how* it will accomplish the strategic goals and mission. It is imperative that all three elements of strategic management be in place to properly set the direction of the organization. Figure 3.3 shows a generic strategic management process flow.

Mission

Some have argued that a company's mission is of little use because big, lofty visions are rarely realized.[4] However, without a mission, a company lacks the overriding statement of purpose for the business (its philosophy toward its customers, employees, and competitors), and its goals such as growth, profitability, or service to the community. Most importantly, without a mission, it becomes difficult to judge the relevance of opportunities, threats, and program options presented to a company's management team.

In our view, defining the company's mission is perhaps one of the most important responsibilities of the senior management team. The mission statement defines why the business exists, what the strategic intent of the company is, what its core values are, and how the management team measures success. It should describe the company's products or services, markets or customers, and specialty areas of emphasis in a way that demonstrates the values, priorities, and goals of the management team.[5]

Financial and economic goals may influence the mission and strategic direction of the enterprise and may be either explicitly or implicitly stated in the company mission statement. Take, for example, the following mission statement from Proctor & Gamble, which explicitly includes sales and profits as part of its mission statement:

> We will provide branded products and services of superior quality and value that improve the lives of the world's consumers. As a result, consumers will reward us with leadership sales, profit, and value creation, allowing our people, our shareholders, and the communities in which we live and work to prosper.[6]

A mission statement for nonprofit organizations is equally important, as they describe why the organization exists, defines the services they perform, and what is driving the passion for human good by the people providing the services. The mission statement for Bella Energy is a good example.

> The Bella Energy mission is to provide the highest quality solar energy solutions to empower communities with opportunities to contribute to a sustainable future and clean energy economy.[7]

A good mission statement promotes a sense of shared expectations among all members of the organization, provides common purpose, and direction for the company, and defines a company's intent for shareholders, donors, employees, customers, suppliers, and the community in which it operates.

Strategic Goals

As stated earlier, the strategic goals define what the company wants to achieve within a multiyear period. To achieve long-term prosperity, management teams commonly establish strategic goals in the following seven areas:[8]

- Profitability
- Competitive Position
- Employee Relations
- Technological Leadership
- Productivity
- Employee Development
- Public Responsibility

In order to accomplish its mission, an enterprise will develop a set of short-, medium-, and long-term goals to focus the effort and outcomes of its workforce. These strategic goals are normally found at two levels: corporate strategic goals and business unit strategic goals.

The purpose of corporate-level strategic goals is to align the various business units toward a common purpose and direction, while business unit strategic goals serve to focus the functional departments and outcomes of the people within a business unit.[9] Every enterprise is unique, and therefore every enterprise will have its own set of strategic goals. It is common, however, to find strategic goals centered in a number of areas, including:[10]

Financial: The ability of an enterprise to achieve its other strategic goals depends greatly on its ability to achieve its financial goals. These goals aim to increase return on assets invested and may include increasing return on investment, increasing profitability or operating budgets, increasing revenue or funding received, or lowering the cost of doing business.

Productivity: Productivity goals normally focus on how efficient an organization is in creating output from input, and may focus on decreasing cost of goods produced, reducing cycle time in creating solutions, or increasing reuse.

Competitive positioning: This refers to a business's relative rank in the marketplace as compared to others that it competes against. This is usually measured in total sales, total revenue, or percentage of total market share.

Customer or user adoption: Adoption refers to the number of customers or users who regularly use a company's offering. These goals center upon increasing the number of people who buy or use an offering.

Improving human condition: Human condition goals are normally focused on creating a positive change in the lives of a select population. Some examples include decreasing incidents of diseases, increasing the use of social services, and changing human rights policies within a region.

It should be pointed out that a business normally does not strive to identify strategic goals in all areas presented above, but rather only in those areas that align with and fully support attainment of the corporate mission. Additionally, a strategic goal should be specific in nature by stating *what* will be achieved, *when* it will be achieved, and *how* it will be measured. A goal must be challenging enough to raise the bar on corporate performance, but attainable to prevent employee frustration and lack of motivation.

Strategies

Strategy defines *how* a business will achieve the strategic goals it has established and consists of the portfolio of ideas that, when fully developed, will contribute to the attainment of those strategic goals.[11] In a program-oriented company, programs constitute the primary strategies that produce superior products, services, infrastructure solutions, and transformative change. The sum of the outputs of the programs within a portfolio will provide the combined means to achieve the desired strategic business goals.

Portfolio Management

A firm's portfolio management process constitutes the second element of the business engine of an enterprise. Portfolio management is defined as "a dynamic decision process, whereby a business' list of active [programs] is constantly updated and revised."[12] As is usually the case, organizations have many more product, service, infrastructure solution, or organizational change ideas than available resources to execute them. As a result, resources can become overcommitted, and the organization must find a way to broker competing demands for its limited resources. Portfolio management is an effective process for identifying and prioritizing programs that best support attainment of the strategic goals. Programs are ranked and prioritized based upon a set of criteria that represents business value to the organization. Senior management can then allocate available resources to the highest value and most strategically significant programs.

According to the model developed by Robert Cooper, Scott Edgett, and Elko Kleinschmidt, the three fundamental principles of portfolio management are: 1) establishing a strong link between the portfolio and organizational strategy, 2) achieving a balance of programs based upon investment priority, and 3) maximizing the value of the portfolio of programs in terms of company goals. We would add that accomplishing these principles should result in matching the number of selected programs with the available resources. This is sometimes referred to as *resource capacity planning*.

Establishing a Strong Link to Strategy

Through the alignment of an organization's investment funding and resources to the work that is the most strategically important, portfolio management maximizes strategic return on investment. As Cooper and his colleagues so eloquently explain, "Well-meaning words on a strategy document or mission statement are meaningless without the funding and resource commitments to back them up."[13]

Ensuring strategic intent can be accomplished through a couple of means. First, the portfolio structure should support the strategic goals established, and second, strategic elements must be included in the portfolio prioritization criteria. Strategically structuring a portfolio can be accomplished by using an *investment bucket* approach. The selection and structure of the investment buckets are unique to each business, and are highly influenced by a company's strategic goals, market segmentation, and product, service, or infrastructure types.

For example, a financial industry software company we worked with defines its portfolio structure on the basis of funding its strategic goals. The strategy statement is as follows: *Drive material impact on (our) margin, growth, and industry influence by bringing to market software solutions and services, and delivering new technologies that enhance the value of (our) solution platforms*.

The portfolio investment buckets for this organization are shown in Figure 3.4. Notice the direct correlation between the strategic intent of the organization and the structure of its portfolio of programs. Also shown in this illustration is the fact that each portfolio bucket consists of a set of programs that become the means to deliver the strategic goals.

Achieving a Balanced Portfolio

Creating a balanced portfolio involves making thoughtful decisions at the macro level on what types of programs the company chooses to invest in, and what level of investment will be provided for each type. This is analogous to the decisions one makes concerning personal financial portfolios. One decides which types of investment vehicles to invest in (stocks, bond, mutual funds, cash, etc.), and what percentage is allocated to each type based on risk tolerance and short-, medium-, and long-term goals.

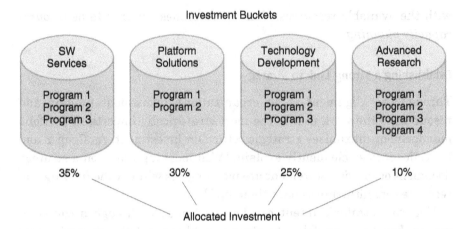

Figure 3.4   Aligning portfolio structure with strategic intent.

For the software company discussed previously, a decision had to be made about the priority of funding and resources to be applied to each of the investment buckets within the portfolio. As shown in Figure 3.4, 35 percent was allocated to software services programs, 30 percent was allocated to platform solution programs, 25 percent was allocated to technology development, and 10 percent was allocated to research.

Balancing the portfolio achieves two critical elements for a business. First, it allocates funding based upon strategic intent and priority. Second, it dedicates funding and resources to specific programs that constitute the portfolio.

Maximizing the Value of the Portfolio

The foundation for an effective portfolio management process is the ability of the management team to determine the factors that constitute program value for their organization. The team must establish a prioritization system to evaluate one program against the others within a portfolio. Once it gets beyond this hurdle the organization will have accomplished the following two things: 1) a real understanding of how value is defined, and 2) a prioritized portfolio of programs that clearly differentiates between those that provide the most and the least value to the business. Once these two are accomplished, maximization of the portfolio has been achieved.

Figure 3.5 is an example of a prioritized portfolio of programs from a manufacturer of Internet communication products. Prioritization of the portfolio is accomplished through the use of two evaluation criteria: 1) net

Program Portfolio	EVALUATION CRITERIA					RANKED PROGRAMS	PRIORITY
	NPV ($M)	Likelihood to Succeed			Risk x NPV	Risk x NPV Ranking	Resource Priority
		TECH	MRKT	Overall			
Program A	$38	20%	70%	14%	$ 5.3	Project D	1
Program B	$15	15%	50%	8%	$ 1.2	Project F	1
Program C	$13	35%	75%	26%	$ 3.4	Project E	1
Program D	$13	80%	80%	64%	$ 8.3	Project A	2
Program E	$19	45%	70%	32%	$ 6.0	Project C	2
Program F	$9	90%	85%	77%	$ 7.0	Project H	2
Program G	$9	50%	45%	23%	$ 2.0	Project G	3
Program H	$5	95%	60%	57%	$ 2.9	Project B	3

NPV = Net Present Value

Figure 3.5 A portfolio of programs ranked by value and risk.

present value of each program, and 2) the combination of technology and market adoption risk. Each program within the portfolio was first ranked based on net present value. The net present value of each program was then multiplied by the overall risk (technology risk × market risk = overall risk) associated with each program. The programs were again ranked based upon the risk multiplied by NPV. Notice how the prioritization changed with the use of a second criterion.

Once the ranking was known, allocation of resources then occurred in priority order, with priority one programs receiving full allocation of resources, priority two programs receiving full allocation of resources once priority one programs were fully staffed, and priority three programs receiving full resources once priorities one and two programs were fully staffed.

With programs ranked and resource prioritization complete, the management team of the organization was able to allocate its resources to the programs that provide the most business value to the organization, therefore achieving maximum output from a limited input.

Executing good decisions that give the portfolio a robust set of programs that reinforce the strategic goals of the business requires strong leadership from executive management. Executives must ensure that the process is effective and that the tough choices are objectively made through quantitative and qualitative analysis, dialogue, and debate. Without strong leadership, an organization will continue to overcommit its resources and underachieve attainment of its strategic goals. The example entitled "Do You Agree with This Portfolio Verdict?" demonstrates one such scenario.

Do You Agree with This Portfolio Verdict?

Peter Bell is the Research and Development Director for StarTech Software, a software application development company located in Ottawa, Canada. Bell's NPI (New Product Introduction) programs were taking far too long to complete. His initial assessment of the problem was that the NPI pipeline was clogged, with too many programs trying to move through it—twenty eight in total. This was causing excess demand on Bell's limited resources, particularly his software coders.

A review of the NPI portfolio was revealing. There were some obvious good programs, but Bell was perplexed by the number of lower-value programs in the portfolio. Additionally, no two programs were in the same market segment area. The approach seemed like a scatter gram of programs with no overriding strategy to guide program evaluation and selection. The lack of strategic focus was evident.

All of the programs held a promise of high reward, which was obviously a good sign. The problem was that they were all high-risk, with a reasonable probability of failure, either technically or commercially. Additionally, almost all programs were long-term. The absence of *quick hits* to balance the long-term development programs to bring in near-term revenue was apparent. Also, almost all of the programs were in the programming phase, with no programs in the requirements definition or closure phase.

"Okay," Bell thought to himself. "Now the issues seem clear. In order to talk to my direct reports, I need to summarize this into a few words to describe the current state of our portfolio of programs." This is how Bell summarized his findings:

- Our portfolio is not focused on our business strategy!
- Our portfolio is poorly balanced!
- We are taking on too much business risk!
- There are too many programs, causing resources to be spread too thinly!

Do you agree with Peter's assessment?

PROGRAM MANAGEMENT AS PART OF THE BUSINESS ENGINE

The role of program management in a firm's business engine is to be the delivery mechanism of the strategic goals established during the strategic management process and for achieving the business benefits identified during the portfolio management process.

An organization's program managers *do not* create the mission or strategic goals; this is the role of senior management. The program managers *do not* plan and execute the project deliverables; this is the role of the project teams. The program managers *do* ensure the attainment of the value proposition of a program by delivering an integrated solution through the collaboration and coordination of multiple interdependent projects and business functions. This ensures alignment between business strategy and project execution is established.

But what is alignment? Alignment refers to the degree to which a program mirrors and supports the priorities of the organization's business strategy. Additionally, it is the degree to which the business strategy is used to guide objectives and work outcomes of the execution team. The program vision is the keystone element that provides this alignment. It establishes the end state that defines success for the program and provides guidelines for what to do and how to do it. The program vision is composed of three critical elements: 1) the program objectives, 2) the whole solution, and 3) the program business case.

Program Objectives

Remember, each program delivers a portion of the overall strategic goals for a firm. It is rare that the successful execution of a *single program* results in the attainment of all strategic goals. Rather, it takes the successful execution of a number of programs within the portfolio.

Each program then carries a set of program objectives that are designed to achieve specific strategic goals. As Figure 3.6 illustrates, the program objectives result from a combination of the various strategic goals of the organization.

The program objectives provide the translation from strategic business goals to actionable execution objectives specific to a program. For example, a strategic business goal to achieve increased market growth could be translated into a program objective to reach a specific number of orders in the first year following capability introduction. Similarly, a strategic business goal to provide lowest-price offerings could be translated into a well-defined cost objective for a program.

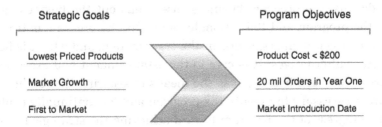

Strategic Goals	Program Objectives
Lowest Priced Products	Product Cost < $200
Market Growth	20 mil Orders in Year One
First to Market	Market Introduction Date

Figure 3.6   From strategic goals to program objectives.

The Whole Solution

When we are working with program teams to establish a strong program vision, we use the term *whole solution* to describe the product, service, infrastructure solution, or transition that the program team will create and introduce into the market or organization. As we explain in more detail in Chapter 4, the whole solution includes the conceptual architecture of the solution and all other components that have to be created or enabled to completely meet the expectation of the customers and users.

This tends to be one of the most difficult tasks in creating the program vision, as it requires a team to visually represent the whole solution that they and their partners will provide. The level of difficulty is matched only by its level of importance as it is the whole solution that defines the program architecture and structure of the program team and demonstrates the level of cross-project integration that needs to occur to achieve the business goals driving the program. When completed effectively, the whole solution shows that success cannot be fulfilled by any one specialist or set of specialists on the team. Rather, success comes when meeting customer expectations is a shared responsibility between the members of the project teams, with their work tightly interwoven and driven toward the integrated solution.[14]

The Program Business Case

The purpose of the program business case is to demonstrate that a program supports the strategic goals of the organization and is used to determine if the organization should invest the financial, human, and capital resources to fully execute the program (Chapter 6). Portions of the program business case are vital in establishing the program vision by describing the business opportunity available and how the program will achieve the opportunity and business strategy.

To do this, the program business case spells out the business benefits of the program and the rationale as to why the program outcome is desired by the customers, users, or the organization, and why it is better than other alternatives. It is core to the establishment and execution of the program vision because it is the means to securing the funding and resources necessary to execute the program and for continually evaluating the progress of the program toward achieving the strategic business goals intended.

In program-oriented organizations, development of the program business case is led by a program manager and provides the direct link between the program and the firm's portfolio process. In these organizations, the approval of a program business case is the means for a program to enter the portfolio. As explained earlier, the program is then evaluated against other programs vying for organizational funding and resources. Clearly, the stronger the business case, the higher the probability a program will receive the investment needed.

THE EXECUTION ENGINE

The execution engine of a company is responsible for turning ideas into reality, turning architectures into solutions, and turning strategy into action. It is widely recognized that strong and healthy companies have strong execution engines, as this is where the means for generating growth and positive change resides. The execution engine of the integrated management system comprises three primary functions: 1) team execution, 2) project management, and 3) program management.

Team Execution

At the foundation of the execution engine are the teams of functional specialists whose knowledge, skills, and expertise are honed for creating new capabilities for the enterprise, its customers, and its stakeholders. Specialization of labor has long been recognized as the keystone to advancements gained during the industrial and modern ages.[15] Because programs tend to be complex in nature, the solutions that are created from the work performed on the program need to be systematically organized into functional specialties. This is core to a systems approach.

It is common for execution specialists to report to and be part of functional departments within a company that serves the purpose of maximizing the competency of the specialty. The functional departments then assign their specialists to the programs as needed. Within program-oriented organizations, the functional specialists become part of the functional- or discipline-specific project teams where the charter of the project is to create and deliver the subsystems that make up the whole solution. Their work therefore is guided by the objectives of the program as a whole, but through the execution of the projects within the program.

When the integrated management system is operating effectively, each member of an execution team should understand how their work contributes to both the creation of the integrated solution and achievement of the business goals desired. If done correctly, a software engineer, for example, will be thinking about business value as well as about the performance of the code he or she is creating.

Project Management

The strength of a company's execution engine is its ability to become highly efficient. This is accomplished through repeatability and predictability of every process, every task, and every activity—this is the charter of the project management function within program-oriented organizations.

The Project Management Institute's PMBOK® provides a masterful guide of effective methods for driving projects to success in the modern world. The project management knowledge areas and processes contained in the PMBOK® are designed to complete work in the most efficient, repeatable, and predictable means possible. The application of these practices is as relevant on a program as they are on an independent project. However, management of a project within a program involves some significant differences.

The first, and debatably the most important, is a difference in mindset. The overall focus for a project team must shift from one of *my* project, to a mindset of *our* program. Thus, a duality is introduced on projects within a program where a project manager must maintain command and control of a project, while collaborating with all other actors on a program to achieve program-level success. This is due to the highly interdependent nature of the projects on a program.

Since projects are evaluated on program-specific success factors, project managers must define, scope, and plan their projects in relationship to the overall program. Project scope is not defined entirely by a project's outcomes and deliverables, but rather also by its contribution to the program objectives and business benefits intended. This requires that the project managers within a program directly link their project deliverables to the stated program objectives. One must be able to trace deliverables to project success criteria, to program objectives, and to business benefits.

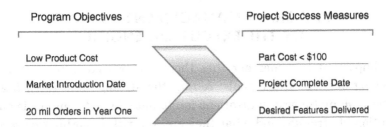

Figure 3.7   From program objectives to project success measures.

Just as a program delivers a portion of the overall strategic goals for a firm, each project on a program delivers a *portion* of the program objectives. As the example in Figure 3.7 illustrates, the project performance measures are driven by the various program objectives.

For example, a program objective of *low product cost* could be translated into project success criteria such as *part cost < $100* as well as *total budget at completion* for a number of the projects on a program. Likewise, a *market introduction date* program objective will likely translate into *project completion date* success criteria for each of the projects.

The alignment of project success criteria to program objectives requires project managers within program-oriented organizations to evolve from being strictly command-and-control execution managers to also performing certain aspects of the role of a first-level business manager. This is necessary to help ensure that the project outcomes from the execution engine of a firm are driving toward business benefit achievement.

Project managers on a program have to define their projects in terms of business benefits and as a single component of the whole solution. Project managers therefore have to be able to think systematically and holistically about their outcomes and deliverables, and be able to tailor their project performance indicators toward achievement of program objectives, not toward project performance entirely.

This evolution requires some change to the traditional project manager role. Seasoned project managers have years of experience being held accountable for the success of their own independent projects. As a result they have honed their philosophies, knowledge, and skills toward becoming masters at managing to baselines, completion of project-focused tasks, deliverables, and milestones. Moving from a philosophy of individual accountability in managing an independent project to a team accountability philosophy necessary on a program can be a challenging transition for some organizations and will take some time and diligence.

PROGRAM MANAGEMENT AS PART
OF THE EXECUTION ENGINE

It is important to remember that there are two parts to successfully achieving one's strategic goals: *setting* the strategy and *executing* the strategy. Many times organizations fail to put adequate emphasis on the execution of strategy, and what they end up with is nothing more than an unfulfilled strategy statement.

The role of program management in a firm's execution engine is to provide translation of business activities to execution activities, perform the integration and synchronization of outcomes from multiple projects, and help the project execution teams effectively and efficiently navigate market and organizational change.

Business to Execution Translation

When gaps occur between a firm's execution activities and business activities, many times it is due to ineffective translation between the two types of activities. Program management provides this critical role. For example, program management translates strategic business goals to program objectives as described previously. As part of execution, the translation must continue by translating program objectives to key project performance and success measures. This ensures that the direction, command, and control of each project is aligned to and supports the strategic goals driving the program as a whole by creating traceability from strategic goals, to program objectives, to project performance.

Additionally, the program management function provides the translation of the whole solution to the program architecture that defines the projects and other components that are needed to create and deliver the whole solution. Additional translation is needed for the execution teams to convert program components to project outcomes or deliverables. This is normally accomplished through the development of a program-level work breakdown structure (Chapters 6 and 9).

Cross-Project Integration and Synchronization

The second critical role that program management fulfills for the execution engine is that of integrating the outcomes of each of the projects within the program to create and deliver an integrated solution.

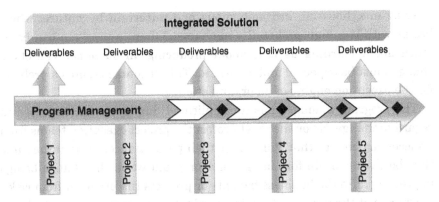

Figure 3.8   Cross-project synchronization and integration.

Figure 3.8 illustrates how program management provides the systems-level synchronization, management, and integration needed to tie the components of a solution together to create a holistic outcome. In doing so, program managers works horizontally to drive collaboration *across* the project specialties.

Integration of project outcomes and deliverables is impossible if the work within the projects is not occurring at a synchronous cadence at the program level. In this regard, a program manager is much like an orchestra conductor. Even though each of the instrument sections of an orchestra has their own music to produce, the conductor ensures each of the sections steps through the musical composition at a consistent tempo and in concert with each of the other sections. This ensures that an integrated, blended, and harmonious musical piece is produced.

Much is the same on a program. Since the responsibility for managing the interdependencies between projects on a program falls upon the program manager, he or she must work to ensure that the workflow within each project is occurring at an integrated and harmonious pace. Synchronization, therefore, involves ensuring that project timelines are aligned, that cross-project deliverables are planned appropriately, and that the work within each project is occurring at the appropriate pace.

Navigating Change

Execution teams by nature and by design resist change. Project success is determined by evaluation of the amount of variation from a preset baseline. Change inherently introduces variation from the baseline, and therefore introduces risk to execution success.

Programs, however, are ambiguous and uncertain by nature, which brings with it a high probability of change. Often, a program can be as much about learning as it is about producing an outcome. Therefore, change must be expected and embraced. This of course creates a problem for the execution engine of an organization.

The program manager has a complicated duality to contend with in regard to change. He or she must protect the project execution teams from unnecessary change that will disrupt their performance, while at the same time be the advocate for change on a program when the change brings improvement to the business benefits or protects the program from risk.

Change at the program level is a *navigation* process, not a control process. Since a program is ambiguous and uncertain, project execution—and the program in its entirety—cannot be planned and executed to a strict baseline and set of targets. This is a hard lesson learned that we have witnessed many organizations struggle to comprehend, especially those early in the process of transitioning from a predominately project-oriented organization to being more program-oriented.

To be successful in a changing environment, the program as a whole and the execution team specifically must be given a set of thresholds that act as guard rails around their baselines. Additionally, buffers must be established for time and cost—like upper and lower control limits or parameters of acceptable performance. This will allow the program team latitude to navigate change successfully instead of attempting to control change which is nearly impossible.

ALIGNING EXECUTION WITH STRATEGY

For program-oriented companies, it is not by coincidence that program management is at the heart of both the business and execution engines of the enterprise—it is by design. This is not to say, however, that program management is the most critical of a company's management systems. All are important and each has its function to serve. For program management that function largely is to create and maintain alignment between execution work outcomes and what the company is trying to achieve strategically.

As we stated earlier, many companies perform exceptionally well without adopting program management. Many times the existence of an effective portfolio management process is sufficient to align a set of independent projects to the strategic goals of an organization. However, the need for transition to something different is triggered when the organization begins to experience a rise in complexity. This may include complexity of the solutions under development, complexity associated with the number of business functions required to be involved in creating solutions, complexity associated with the geographical distribution of work, and complexity associated with strategic alliance and co-development partnership arrangements.

The result of this rise in complexity is often realized in a misalignment between project deliverables and the business strategies within the organization. This misalignment eventually causes disruption in business results. In working with a number of firms, we have witnessed a threshold of complexity that, once reached, triggers the beginning of the disruption. For project-oriented organizations, disruption may occur because of a non-incremental increase in the number of projects required to manage the complexity, and the need to manage a large number of cross-project interdependencies.

For a number of years, some authors were advocating the need for *strategic project management*, where they stressed the importance for project managers to develop strategic business skills in order to fill the gap that was arising. While good in theory, in practice this makes little sense. As stated earlier, strategy has two sides, development of strategy and execution of strategy. Project management is best when hyper-focused on effective and efficient execution of the strategy. Becoming more "strategic" does little more than serve to dilute the value of project management for an organization.

Rather, we have found that senior leaders of organizations begin to look for a more systematic and integrative approach to solve the disruption problem caused by increased complexity, and begin to explore the principles of program management as a solution. Here is what one such leader, David Churchill, a former Vice-President for Agilent Technologies, said about the role of program management:

> Most firms have the right product ideas, technical talent and marketing capability to support their business strategies. Organizations many times have difficulty performing to expectations when they cannot turn their strategies into successful execution. Program management is strategic to the firm because it provides the ability to convert business plans into actions that will achieve the intended objectives; it helps bridge the gap between strategy and execution.

With this, however, is the challenge to establish program management correctly within an organization, and to focus on the critical aspects of program management that creates the strategy-execution alignment. The critical aspects that are presented in this chapter are a strong program vision and documented traceability between strategy, program, and project success measures. Table 3.1 illustrates a number of examples that demonstrate the alignment between strategic goals, program objectives, and project success measures.

It is critical to remember that alignment is not static and should not be treated as a one-time activity. Aligning strategic goals, program objectives, and project performance measures are normally performed early in the program cycle, at a time when much is still unknown and many assumptions are made about the environment in which the program

Table 3.1   Strategy, program, and project alignment examples.

Strategic Goals	Program Objectives	Project Measures
Be first to market with a new web-based service	1. Market introduction date 2. Highest customer satisfaction rating	1a. Performance against schedule 2a. Product feature performance
Provide the lowest priced product on the market	1. Low product cost 2. Low development cost	1a. Bill of material cost 2a. Performance against budget 2b. Percentage of features developed in software versus hardware
Increase throughput by decreasing time of customer transaction and quality of service	1. "X"% decrease in customer service time 2. Increase quality of service	1a. Reduction of customer service steps 2a. Resolution of customer-identified "pain points" 2b. Customer "call-back" occurrence reduction
Meet demand for a 25% increase in services by 2017	1. 22,000-square-foot building expansion 2. Completion date by August 2017 3. 50% increase in service content 4. $2 million total cost	1a. Approval of building design 2a. Performance to schedule 3a. Completion of content deliverables 4a. Performance to budget

is operating. The likelihood of change is therefore high and should be planned for accordingly.

The program manager has the responsibility for periodically validating and, if necessary, adjusting the alignment parameters to match market or business changes. Occasionally, this may lead to a discussion with the senior leaders of an organization about the validity of the strategic goals driving a program. This process provides strategic feedback and helps to keep a company strategically aligned to changes in their environment.[16] An example of what can happen when the strategic feedback sensors are not in place is provided in the story entitled "Asleep at the Wheel."

Asleep at the Wheel

Conventional wisdom says that when a program is well aligned with business strategy, it will have a higher probability of being successful. The following true story demonstrates that alignment is not a one-time activity.

The Tripdale program was positioned for success. The program vision, especially the two elements of the whole solution and the business case, were meticulously analyzed and detailed. The product feature set and market price were both chosen in direct negotiations with the lead customer. Detailed program planning was conducted as it had always been—with a team of project-oriented experts. The phase-gate reviews were performed thoroughly and in a timely manner, but without customer representation. The strategic alignment of the program was frequently discussed in program reviews, and with the vice president in charge of new product development. All parties agreed that it was a well-aligned program and on its way to becoming a money-maker.

However, near the end of the program, the lead customers were contacted to announce the program completion and product availability. Customer meetings were scheduled and product reviews were planned. At this time, the project team felt successful—their timelines were met, costs were managed to within the $1 million budget allotment, and they were excited to unveil the product to the customer.

The customers reviewed the product details and stated that they were looking for a different feature set, one that provided more mobility, not greater speed (an apparent strategic shift). What a surprise—the product suddenly had no market! What led to this situation?

During the two-year product development cycle, no regular, periodic contact with the customers occurred. Had this occurred, the program manager would have learned that the market needs had shifted from greater performance to greater mobility. Thus, the business strategy and program alignment had become invalid at some point, and the program team did not know that they had become misaligned. The impact was an $800,000 loss of product development funding!

This could have been avoided if the program team was in frequent contact with customers and end users, and used the information gathered from the customer contact to review program alignment during the periodic phase-gate reviews.

This story brings out a subtle but important point. Many times the program manager is in the best position to identify changes in the business environment. As such, a program manager is then in a position to inform the organization that a reevaluation of the firm's strategic goals and strategies is necessary.

Program Management in Practice: *Keytron Goes Global*

As Scott Jones hung up the phone, the reality of his situation began to set in—there are industry leaders and there are industry followers, and he was working for a follower. Both he and the company he now works for are feeling the pressure to rapidly catch up competitively to the changes within their industry.

Several months prior, Jones was offered and had accepted the Director of New Product Development position with a company in the digital projection industry named Keytron. Although he was not looking to change employers, he felt his career had reached a plateau, and a new challenge would be welcomed if the right opportunity presented itself. At the time, the position at Keytron seemed to be that right opportunity. Now, however, he was not so sure.

Even though Keytron is not one of the top enterprises in the digital projection industry, they have grown at a consistent and healthy rate and have plans to continue increasing their market presence. Primarily through a strategy of mergers and acquisitions, they are positioned to be one of the world's top five digital projection companies within the next five to seven years; that is, if changes are made to increase offerings, increase consistency, and decrease time-to-market performance.

As Director of New Product Development, Jones sees that he will be responsible for this improved performance and he will be at the heart of the company's engine of growth as they create innovative products for new and expanding markets. However, this is a very different and more complex environment than what Jones has experienced to date. As a result of the mergers and acquisitions strategy, new product development at Keytron is now distributed across several countries and continents. Product design is performed at five sites—two in Europe, two in the

United States, and one in India. Additionally, product integration and testing occurs in Mexico, final production is in the process of being transferred from the United States to Taiwan, and a major component for many of Keytron's products is designed and manufactured by a strategic alliance partner in Korea. Jones's new product development teams are now highly distributed across the globe and thus face much more complexity and many more challenges than Jones originally envisioned.

As the person responsible for improving Keytron's development performance in a highly competitive global environment Jones realized he was in a predicament. This highly distributed model is entirely new to him. New product development at his previous employer was performed at a single site in the United States and the new product development teams were co-located and highly integrated. Because of this, he lacks the direct experience in leading distributed teams from which to draw upon and use in his new role. Like any good senior leader, however, he began looking to his network of people who he knew possessed the direct experience in leading global teams for advice, guidance, and support.

Playing a Game of Catch-up

Jones decided to first contact one of his industry colleagues who has experience working for a large multi-national company. Like Jones, Melissa Doyle was a new product development director and worked for a leading manufacturer in the high-tech industry. As Jones and Doyle began their conversation, Jones explained that he had taken a new position since they met last, and was contacting Doyle in hopes of gaining some best practice advice on how to effectively lead global teams.

Doyle congratulated Jones on his new position and gave his inquiry careful consideration. After several minutes of nonspecific conversation, Doyle responded that she honestly could not pinpoint her company's best practices for leading global programs. She certainly agreed that her company was a leader in globalization strategy and execution, but she could not identify the handful of things that constitute their leadership position. She went on to explain that working in highly distributed teams was just how they did things at her company, and it is how they have been creating their products for over 20 years. Distributed teams are merely a component of their company culture, Doyle concluded.

With many more questions raised by Jones, the conversation with Doyle spanned a couple hours. From the conversation, Jones began to understand what it means to be a leading global company. He was able to summarize his learnings in three key points:

1. Global leaders are able to execute their programs and projects in a globally distributed environment as effectively as their toughest competitors do in a domestic environment.
2. Global leaders consistently achieve their strategic goals as a result of a tight alignment between strategy and execution output.
3. Global leaders use global execution as a competitive advantage.

This conversation led Jones to realize that Keytron was an industry follower, not a leader, and that he and his company were playing a game of catch-up within the digital projection industry. Jones realized that as his company began the process of growing globally they were moving away from a comfortable position within a niche market segment of their industry, to a position that would make them a direct competitor to the industry leaders—all of which are *global* leaders.

Even though they were now operating in a global environment, the company's business processes, tools, and organizational and team structures were still based on a local development model in which much of the work was performed at a single site and within a common business and country culture. Additionally, the program managers working for Jones were struggling to overcome the cultural, communication, time zone, and virtual team challenges they were now facing. In short, what was once a rather simple practice of development has now become a quite complex endeavor. Figure 3.9 illustrates the growing complexity encountered when moving from a collocated organization to a highly distributed organization.

As a result, Keytron's development teams seemed to be in disarray. Poor cross-team communication was causing severe development delays; poor documentation and disjointed work hand-offs were resulting in mistakes and rework; and a breakdown in trust had occurred due to missed deliverables and goals. Additionally, organizational, functional, and geographic silos were causing a focus on local solutions instead of a single globally integrated solution. To make matters worse, an investment in a suite of new software-based collaborative tools failed to improve the situation, and was actually making it worse by diverting attention away from some of the core problems causing the poor global execution. Needless to say, Keytron's product development performance was at risk and therefore

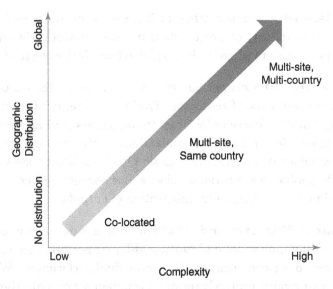

Figure 3.9 Complexity grows with geographic distribution.

their strategic goal to be one of the world's top digital projection companies was also in jeopardy despite their growth strategy.

Scott Jones Is Not Alone

Jones's predicament is not an isolated case. Like Jones, program managers often find that they now have to lead teams that are distributed across the globe. Unfortunately, many of them are not equipped or prepared to deal with some of the significant barriers and challenges they encounter.

As an example, one of Jones's program managers did an outstanding job of stewarding the first version of a product through the manufacturing process and Jones decided she was deserving of special recognition and reward. He chose to publicly recognize the program manager on the company website and in a department meeting. "I did not yet understand that this very typical recognition that I had used many times for United States-based teams, could be quite inappropriate for globally distributed teams," explained Jones. "Since this person was from a collective culture, to be singled out for credit in front of her entire team, each of whom also had contributed to the team's success, was not a rewarding experience at all. In fact, this gesture was a humiliating and demotivating experience for her."

Jones learned a number other of lessons as he navigated his way through the complexity of globally distributed team work. He specifically noted several barriers when leading a global team including the following:

Organizational Structure: Success depends upon effective collaboration among a network of resources. Traditional hierarchical structures create collaboration barriers for program managers leading a highly distributed team. Best practices organizations flatten the organization to increase collaboration. A flat organization leads to direct access to senior leaders, deemphasizes functional silos, shifts decision power to the program, and facilitates horizontal integration of outputs.

Performance Measures and Rewards: Success depends upon a shift from rewarding individual and functional-based performance to rewarding team-based performance as well as individual performance. When people are working collaboratively toward a common set of goals, they cannot be rewarded for competing against one another. Best practices companies emphasize and reinforce team performance.

Development Model: Success depends upon the adoption of a collaborative development model. Traditional project hand-off models do not scale in a highly distributed context. Best practices organizations use a program management model to establish continual alignment of execution outputs to business goals, establish a modular (more flexible) approach, and integrate distributed project work into a whole solution.

Leadership Skills and Competencies: Competencies change as the team becomes increasingly distributed. Best practices companies support program manager training and professional development in leading virtual, multi-cultural teams with the use of internal training, external seminars and courses, rotational assignments, mentoring relationships, and communities of practice.

Scott Jones Three Years Later

It has been nearly three years since the day Scott Jones realized that the company that he was working for was an industry follower in the digital projection industry. Today, however, he is more upbeat about Keytron's prospects of one day becoming a leading company within the industry. This is because he has seen significant progress toward the alignment of the firm's global execution abilities with their globalization strategy. In fact, performance measures such as on-time delivery and increased market share in emerging markets continue to improve business quarter by quarter.

According to Jones, "I have learned many things about program management in a global environment, but one of the key lessons is that it takes time and considerable effort to begin to realize your globalization goals." To their credit, Jones and Keytron at large have made significant progress in their globalization journey. "When I look back, I can see we've made some significant changes," explained Jones. "Our development organization is now more collaborative since we flattened the organization a bit to bring the program teams closer to the senior leaders, the program management model has been well received, our employees are more aware and embracing of our cultural differences, and we're beginning to introduce common development process and tools across the company to increase global team efficiency."

ENDNOTES

1. Bowen, H. K., K. B. Clark, C. A. Holloway, and S. C. Wheelwright. "Development Projects: The Engine of Renewal." *Harvard Business Review* 72(3), 1994, pp. 110–120.
2. Martinelli, Russ, and Jim Waddell. "Aligning Program Management to Business Strategy." *Project Management World Today,* January–February 2005.
3. Pearce, John A., II, and Richard B. Robinson, Jr. *Strategic Management: Formulation, Implementation, and Control.* New York: McGraw-Hill, 2000.
4. Gharajedaghi, Jamsid. *Systems Thinking: Managing Chaos and Complexity.* Woburn, Mass.: Butterworth-Heinmann Publishing, 1999.
5. King, William R., and David L. Cleland. *Strategic Planning and Policy.* New York: Van Nostrand Reinhold, 1978.
6. Proctor & Gamble website, www.pg.com.
7. Bella Energy Website, www.bellaenergy.com.
8. De Wit, B., and Ron Meyer. *Strategy: Process, Content, Context,* 4th ed. London: Cengage Learning, 2010.
9. Pearce and Robinson, *Strategic Management.*
10. Martinelli and Waddell, "Aligning Program Management."
11. Cooper, Robert G., Scott J. Edgett, and Elko J. Kleinschmidt. *Portfolio Management for New Products.* Cambridge, Mass.: Perseus Publishing, 2001, p. 3.

12. Ibid., pp. 140–141.

13. Martinelli, Russ J., Tim J. Rahschulte, and James M. Waddell. *Leading Global Project Teams: The New Leadership Challenge*. Toronto: Multi-Media Publications, 2007.

14. Govindarajan, Vijay, and Chris Trimble. *The Other Side of Innovation*. Boston, Mass.: Harvard Business School Publishing, 2010.

15. Aronson, Z. H., et al. "Project Spirit—A Strategic Concept". International Conference on Management of Engineering and Technology. Portland, OR, 2001.

16. Mintzberg, Henry. *The Rise and Fall of Strategic Planning*. New York: The Free Press, 2004.

Part II

Delivering the Whole Solution

While Part I dealt with the strategic purpose of program management, Part II delves into details of how to manage a program from its inception to end of life. It consists of three chapters: Chapter 4, The Whole Solution; Chapter 5, The Integrated Program Team; and Chapter 6, Managing the Program.

Chapter 4 demonstrates how systems thinking is used as a way to simplify the level of complexity associated with many programs. To do so, we first explain the pressures and sources of complexity that are causing challenges within organizations and then introduce the *whole solution* concept for structuring a program into a set of interdependent projects and project enablers that can be used to satisfy customers and markets.

Chapter 5 continues the systems approach by exploring how to create an integrated program team that can effectively execute the program and deliver an integrated solution. Without careful consideration of how an organization's programs are structured, many or all of the benefits of program management will be unrealized. Having said this, our research has revealed that one of the most common errors businesses make in implementing program management is failing to understand the difference between a program team structure and a project team structure. Therefore, we detail the difference and then examine the process involved with forming an integrated program team that facilitates cross-organizational collaboration required for effective program management.

Chapter 6 wraps up Part II by describing the process of managing a program. As we have learned, management of a program is often a difficult undertaking due to the ambiguity and the level of uncertainty associated with the environment in which a program exists. To effectively manage a program, structure is needed to guide a program's journey from strategy development to benefits delivery. Therefore, in this chapter we detail a business decision framework and discuss the critical aspects of managing a program within that framework.

Chapter 4

The Whole Solution

When does a project become so big that it becomes a program? This is a question that has been posed to us on a number of occasions. Unfortunately, we do not have a good answer for this question as we view things a bit differently. Our experience is that it's not about whether a project should be restructured as a program because of its *size*, but rather because of the level of *complexity* that is involved.

Richard Cook, the deputy project manager of the Mars Science Laboratory at NASA, knows something about complexity. Before overseeing the operations of the Mars Exploration Rover, Cook was the manager of the Mars Pathfinder Mission and before that conducted trajectory designs on the Magellan Project. After years of working with complexity, he concluded that the word *complexity* "is frequently thrown around as a sort of synonym for 'difficult."[1] Cook noted correctly, "Complexity is the quality of being intricately combined," and he distinguished complexity from difficulty based on "the number of interconnected elements that are tied together either technically or programmatically."

It is this perspective of interconnectedness and interdependence among parts that, for us, is the primary determining factor if a project should be structured and managed as a program. The goal is to manage the amount of complexity that develops from the introduction of interdependencies. For decades, people working to create and deliver complex solutions in industries such as aerospace and automotive have used a systems approach to simplify the levels of complexity they encounter. Along with this, program management has been deployed as an effective means for managing complex work efforts from a systems perspective.

This approach is as relevant and effective today as it has been over the past six decades. The difference, however, is that the level of complexity is no longer contained to the historic industries mentioned above. Rather, complexity now permeates nearly all industries and is a key challenge within both public and private sectors as well as within both for-profit and nonprofit organizations.

In this chapter, we demonstrate how systems thinking is used as a way to simplify the level of complexity involved in the development of new capabilities. To do so, we first explain the pressures and sources of complexity that are causing challenges within organizations today. We then introduce the whole solution concept for structuring a program into a set of interdependent projects and project enablers, and explain how companies are utilizing it to satisfy customers and markets.

COMPLEXITY RISING

Complexity, normally referred to as the state of something that has many interconnected parts and interrelationships, is part of our reality. In many aspects of life, humans have a tendency to push the norm or current status quo. Our ever-increasing wants and desires drive our collective environment toward more challenging and exciting ends. This is true in our relationships, activities, careers, and especially in the products, services, and other capabilities we utilize.[2]

Increasingly, we as consumers are demanding more complex solutions driven by specialization and customization demands to meet our individual needs and desires. This complexity manifests itself in the following ways: designs have become more complex as the desire for more integrated features increases; the development of solutions has become more complex by the distribution of resources across the planet; and new innovation has to be married with desirable user experiences.

This push for more exciting solutions has moved us to a world of more connectedness, which is increasingly leading to the pervasive rise in complexity. Things that were once independent (think of phones, cameras, and computers) are no longer so, and the neat and simple construct of the past is no longer viable.[3] For people like Richard Cook who work in industries in which complexity has always been central to the way they do business, this is not a new phenomenon. For many others, this rise in complexity has emerged as a primary challenge to the way we have historically operated within our businesses and organizations.

Recently, the Economist Intelligence Unit (EIU)—an independent business within the Economist Group that provides intelligence and insights on business operations and markets—conducted a survey among business leaders to ascertain the challenge of complexity.[4] Three hundred executives were surveyed and the conclusions were as follows:

- Doing business has become increasingly complex.
- The biggest cause of complexity is customer expectations.
- Businesses are finding it increasingly difficult to cope with this rise in complexity.
- Historic business processes and organizational structures are adding to the complexity challenge.

As noted by the EIU research, one of the major causes of rising complexity is the result of customer expectations, and our historic ways of doing business have become insufficient to manage the new levels of complexity.

While speaking at a conference on the topic of complexity and systems thinking, we were approached afterward by a gentleman who seemed to relate to what we were espousing. "This is exactly what we need," he stated. "We are being consumed by complexity." What he went on to describe was a situation where the infrastructure technology capabilities that were being developed and deployed within his organization (a well-known banking institution) had become so complex that they could no longer be effectively managed as a single project.

He explained that the firm attempted a multi-project approach for the development of their capabilities by sub-dividing the effort into multiple projects, each with a dedicated project manager and team. What the gentleman saw in our presentation that made him realize that this approach was insufficient was the large network of cross-project interdependencies and interconnectedness that existed between the projects. These interdependencies were being left unmanaged. The following explores this in more detail.

The interdependencies between projects consist of a series of cross-project deliverables, coordinated tasks, cross-discipline decisions, and shared management of risks and problems that are encountered on any complex work effort.

As the term implies, interdependent projects are those that have a dependence upon the delivery of an output from one project to another. Project interdependencies normally equate to deliverables, tangible

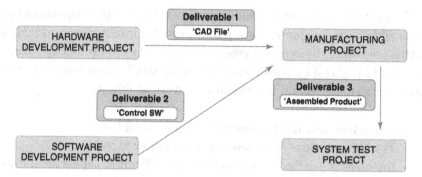

Figure 4.1   Cross-project interdependence.

outputs from one project team that are delivered to one or more other project teams. Interdependency between the project teams is formed when a deliverable from one project team is needed to successfully complete the work of a second (or more) project team(s).

To illustrate how interdependencies between projects are a determining factor in complexity, Figure 4.1 provides a simple example consisting of four projects.

The three deliverables in Figure 4.1 are: 1) the computer-aided design (CAD) files delivered from the hardware project team to the manufacturing project team, 2) the control software delivered by the software project team to the manufacturing project team, and 3) the manufactured product delivered from the manufacturing project team to the system test project team.

This series of deliverables creates a highly interdependent relationship between the projects. For the first deliverable, the manufacturing project team needs the CAD files from the hardware project team for the production of the physical circuit boards. An interdependency is established between the hardware development and manufacturing project teams.

The manufacturing team cannot build a circuit board without the control software from the software project team that is needed to power up and control the board. This, of course, would lead to failure of the system test project team to complete the test of the product.

The three deliverables described create a highly interdependent relationship between the four project teams. These interdependencies between project teams need to be managed as closely as the deliverables created by each project team. Focusing on completion of deliverables alone is not enough to be successful, especially on projects with significant complexity and numerous interdependencies.

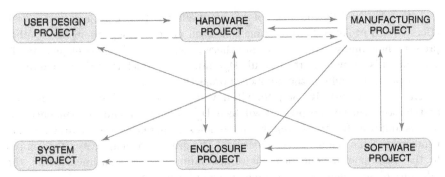

Figure 4.2 Increased complexity with increased interdependence.

Determining the Level of Complexity

The level of complexity is highly dependent upon the number of cross-project interdependencies and should be a key deciding factor of the management approach used to oversee the initiative. To illustrate, we will simply add two elements to the example above. First, we will assume that the product requires the development of an integrated circuit (IC), and second, we will assume that the product requires the development of an enclosure to power, cool, and contain the other components. Our expanded set of projects and dependencies is illustrated in Figure 4.2, with each arrow representing a potential interdependent deliverable between the projects.

One can visualize that the complexity of the interdependency structure between the projects increases with the addition of just *two* elements. In fact, a total of 30 potential interdependencies between the 6 projects are now possible (see "How Many Dependencies Are There among Five Projects?"). One can quickly see that as the magnitude of the effort increases, the number of interdependencies among project teams becomes a management challenge, thus highlighting the need for an effective approach to provide focus on the interdependencies between projects.

How Many Dependencies Are There among Five Projects?

Some organizations use size as the key variable when determining complexity. The reasoning is that as the size increases, so does the number of management activities. Such increased complexity, then, has its penalty—larger work efforts require more managerial work and deliverables to coordinate the increased number of interactions.

Let's take the following examples: a work effort with five projects and a work effort with 6 projects. If we assume that each project is dependent on all other projects, how many dependencies are there in each of the two developments? To calculate this, we will use the formula $D_n = n(n-1)$ in which D_n is the total number of dependencies and n is the number of projects.

The scenario with five projects will have $D_n = 5(5-1) = 20$ interdependencies, and the scenario with six projects will have $D_n = 6(6-1) = 30$ interdependencies. It means that from five to six projects, the complexity increases from 20 to 30 interdependencies. This is a 50 percent increase! The growth of complexity (if measured by the number of dependencies) is compounding, rather than linear.[5] And, the difficulty of management compounds in a similar fashion.

One could make the case that the project managers from the four projects illustrated in Figure 4.1 could adequately manage the work effort described on a cooperative basis without the need for a program manager. We do not disagree that this may be true. In fact, projects of this nature are managed effectively every day by capable project managers. However, industry benchmarking, research, and personal experience show us that this approach is sustainable only to fairly low levels of project complexity.

Four serious problems are consistently encountered when complexity overrides traditional project approaches. First, as complexity rises, the project managers are required to spend the majority of their time planning, executing, and delivering their pieces of the work effort. Eventually, the work output has to be integrated, which is rarely a flawless event. The project can experience significant schedule and cost overruns due to rework and extended integration efforts caused by lack of focus on interdependencies.

A second problem is encountered when schedule pressures exist due to internal or external factors. When schedule pressure exists, the natural tendency for project managers is to hunker down and execute their project deliverables, with little or no time allocated to managing the business goals driving the need for the project. Business requirements such as return on investment or cost reduction goals may be left unattended. In this scenario, the capability may achieve all technical features and functional requirements intended and may be delivered to the aggressive schedule target, but the project may also be deemed a failure because the business goals were not achieved.

Problem three is that someone needs to be responsible for delivering the whole solution to the customers and stakeholders. As the scenarios in the first two problems begin to unfold, the project teams begin focusing on their respective elements of the solution exclusively; after all, that is their "job one." Therefore, the technical aspects of each functional discipline take precedence over the integrated solution when, in fact, just the opposite has to occur to adequately meet customers' expectations.

The fourth problem that can be encountered is that interdependencies between projects can grow at a rapid rate as demonstrated previously. Eventually, the cooperative approach by the project managers to manage the mesh of interdependencies breaks down, and no one is left to manage the large number of interdependencies between the projects. This is a critical point. There are two vectors of complexity that need to be managed. The first vector is the definition of responsibilities and deliverables for each individual project manager and his or her project team. The second vector of complexity involves defining, sorting out, and managing the large number of interdependent tasks, responsibilities, and deliverables that require joint sharing of information, planning, and executing *between* the project managers and members of the various project teams.

This nature of interrelationship and collaboration necessary today is ubiquitous and has caused us to shift our focus from the management of individual elements alone, to also focus on a more macro-level approach of managing the *interactions among the elements*. This is the basis of systems thinking.[6]

SYSTEMS THINKING

For the most part, we see the world as increasingly complex and uncertain because we are trying to use inadequate concepts to explain it. In the world of minimal interconnectedness of the past, we were able to employ basic analysis to explain things. First, we isolate the parts we are trying to understand; we then characterize the behavior of each individual element we are studying; and finally we aggregate our understanding of the parts to explain the behavior of the whole.

In the current world of interconnectedness, however, a different way of thinking is required to understand and characterize behavior. Systems thinking does not require separation of the parts from the whole. Rather it respects the interrelationships between the parts. More importantly, it focuses on the *interactions* between the parts.

The systems thinking approach allows one to step above the fray of details associated with complexity to view the world from a simpler vantage point. A recent article in the Wall Street Journal recounted Henry

David Thereau's maxim for "simplicity, simplicity, simplicity."[7] Thoreau was convinced that our lives are "frittered away by details."[8] What he was telling us was that complexity has congested our lives and that the solution to complexity is "simplicity."

Fortunately, people are continuing to come to the understanding that systems thinking is an effective approach to simplification of complexity and uncertainty. George Reed, a retired army colonel, is one of those people. Regarding the need for systems thinking, he noted: "Leaders operate in the realm of bewildering uncertainty and staggering complexity,"[9] and to successfully operate in any contemporary market or organization today requires systems thinking.

Systems

Ludwig von Bertalanffy, an Austrian-born biologist, is recognized as the father of general systems theory and systems thinking. Leveraging von Bertalanffy's work from the 1940s and '50s, a system can be thought of as *a combination of interrelated parts that collectively work together as an integrated whole to achieve a common purpose*.[10] Each part by itself is of little or no value. Value is only created from the interaction with the other parts of the system to perform the function of the whole. Think of an analog clock. The parts of a clock, such as cogs, dials, and hands, are of little value by themselves. They only provide synergistic value (customer determined value) when they are assembled together and collectively perform the function of the clock (the system).

A system can be viewed as a simple entity consisting of four primary elements: 1) inputs, 2) outputs, 3) interdependent subsystems, and 4) an environment within which the system operates. Figure 4.3 illustrates a simple model of a system.

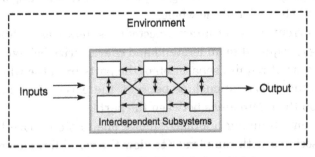

Figure 4.3   A simple systems model.

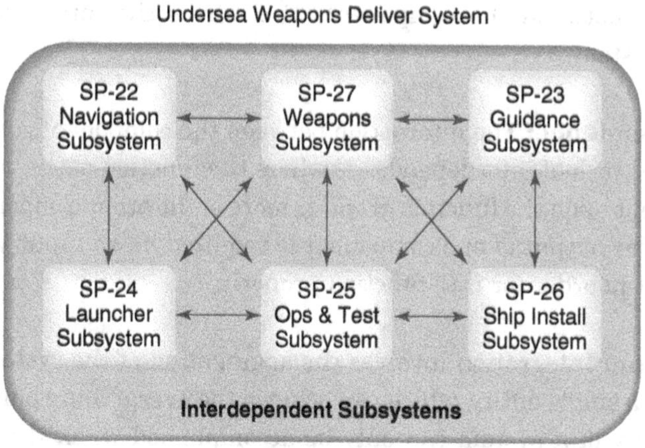

Undersea Weapons Deliver System

Figure 4.4   A simple model for a complex system.

As we shared in Chapter 1, the first documented case of the use of program management in the U.S. dates back to 1957. Within that description also lies an example of a complex system being described by a simple model. The system described was an undersea weapons delivery system (remember, this was the era of the Cold War). Figure 4.4 illustrates the model showing the major components (or subsystems) that comprised the overall system.

The diagram also shows the network of interdependencies between the subsystems. We have greatly simplified this interdependency network for illustration purposes. In reality there were hundreds of interdependencies among the subsystems.

We use this simplified example to make a number of key points: First, complexity in capability development has existed for a long time. Second, systems thinking has successfully been employed as a way to simplify capability complexity. Third, there are a number of key characteristics of a system that are relevant to the way we should manage complex programs today. These characteristics include the following:

Organization: A system is organized in a way that brings structure and order to the components and to the whole. It is the arrangement of the components that helps the system accomplish the overriding purpose or goal.

Interaction: Interaction refers to the manner in which each of the components of a system functions with the other components. It is the interaction between the components that forms the interrelationships within a system.

Interdependence: The interaction between the components of a system forms a network of interdependence where the function of one component is dependent upon the function of one or more of the other components. For example, the output of one component is required as an input to another component in order for it to function properly.

Integration: Integration involves the aggregation of the system components into a single entity, which can achieve the overarching purpose. The functionality of a system can only be accomplished when the activities associated with each of the components are tied together.

Our intent here is not to provide a crash course on systems theory. That is beyond the scope of this book. We provide the information above, however, to make the point that there is great value in taking a top-down, systematic approach to define your next program, especially as the complexity of your program increases.

This approach applies for strategy-led programs and formulated programs, as well as for mandatory programs that are intended to meet regulatory or compliance stipulations. By taking a top-down approach, a program manager will be working to create a critical component of their program—the program vision.

When working with organizations and their program managers, we use the whole solution concept to establish the program vision. You will likely now recognize the value of systems thinking in the discussion that follows.

THE WHOLE SOLUTION CONCEPT

With the exception of a small percentage of companies, most organizations do not and will not utilize a full systems approach to create new capabilities. This does not mean that they cannot benefit from the concepts of systems thinking, particularly if they wish to implement some of the fundamental aspects of program management, such as the importance of taking a top-down approach where one begins with a firm's business goals and the strategies to achieve them as the hinge points for defining and managing a program.

When complex, system-level opportunities arise, a full range of company specialists are needed to create solutions. We have found it useful and effective to use the *whole solution* concept when first introducing systems concepts to an organization to establish the program vision, or the capability end state, that the program is intending to reach.

The concept of the whole solution is not new. It originated from the marketing discipline when Geoffrey A. Moore coined the term "The Whole Product" in his book *Crossing the Chasm*. He defined the whole product as the products and services that best meet the customer's wants and needs (see "The Concept of the Whole Product").[11]

The Concept of the Whole Product

The whole product concept is simple; there are two compelling value propositions for each company's capabilities, as follows: First, the expectation on the part of the customer that their wants and needs be met. Second, the ability of the company to provide a capability that fulfills the wants and needs. Many times there is a gap between the two. To close the gap, the company must add an array of services and ancillary products to the original solution, thereby creating the **whole product**. To understand this, it is best to understand the different forms of products.

- **Generic product** is what is shipped in the box and purchased.
- **Expected product** is the product that meets the *minimum* expectations of the customer.
- **Whole product** is the product providing the *maximum* chance to achieve the customer's buying objective.

In marketing battles, the generic product is the center of the battle for the early market. When the market shifts toward a mainstream market, more sophistication is needed to win, and the center of the market battle shifts to whole product solutions.

If we shift from a marketing focus to a capability delivery focus, the whole solution can be defined as "the integrated solution that fulfills the customers' expectation."[12] In other words, the defined solution must holistically meet all of the customer's expectations.

As many companies and organizations have learned, meeting the customer's expectations becomes the means to achieve the strategic business goals of the firm. If customers put a priority on receiving the whole solution, the concept then needs to be part of a company's business strategy. Once it is part of the business strategy, the program management discipline becomes responsible for developing and delivering the whole solution.

If we purchase a laptop computer, for example, we wouldn't consider it acceptable if we were delivered a box of circuit boards, a second box that contains the enclosure, another box containing peripheral devices such as memory and network adapters, and finally an envelope containing the computer software applications and operating system on a set of compact disks. Rather, unless we're a computer hobbyist or a systems integrator, we *expect* to receive an integrated laptop that we can unpack, plug in, and begin using with the applications we most desire. Thus, we want to experience the delivery of the whole solution.

Delivering the Whole Solution at Nike

The sports giant started in 1964 with the name Blue Ribbon Sports, changing its name to Nike in 1978. Since then, the company has mastered the complex landscape of the sports apparel market (from design to sourcing to manufacturing to delivery). Few understand the value of the whole solution concept better than Nike.

The Nike business model is based on the whole solution concept. It understands the value of efficiency that comes from standardization, yet also knows that customers have individual wants and needs when it comes to using sports apparel and gear in their work and personal lives. This is the reason for NikeiD.

Standard products are fine. Nike has realized that some buyers are happy with basic, generic, off-the-shelf products. A growing number of buyers, however, want greater value by being able to customize their product based on individual needs and preferences. The NikeiD service provides this opportunity.

NikeiD allows the customer to customize Nike products. This service puts the design of the quality Nike products in the hands (and minds) of its customers. The customer is the designer, which adds to the complexity of this market in terms of software to enable the customization (front end) tied into a commerce engine (at Nike) and seamlessly at the manufacturer (back end) to produce the custom product and logistics for delivery.

When one uses the NikeiD service, he or she starts with a standard-version product. From there, the customer redesigns the product color schemes, logo insignia, and adds personalized signage to meet any unique needs and personality preferences. Nike, therefore, provides the whole solution—the integrated solution that meets the customer's expectations.

It is helpful to think of the whole solution consisting of two parts: 1) the core components, and 2) the enabling components. The core components are the tangible elements of the whole solution that, when integrated,

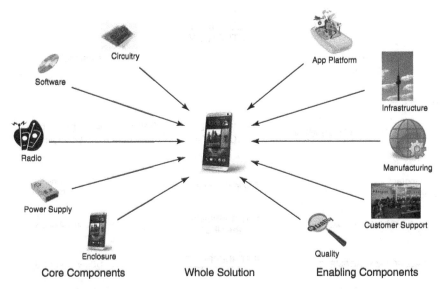

Core Components   Whole Solution   Enabling Components

Figure 4.5 Diagramming a smart phone whole solution.

constitute the physical capability developed. In systems language, these are the subsystems of the integrated system. The enabling components are the additional elements needed to ensure the capability meets customers' expectations.

Let's look at an example to illustrate. Imagine that you are in charge of leading the program commissioned to create the next generation smart phone for a leading phone manufacturer. Your whole solution diagram may look something like the one illustrated in Figure 4.5.

The whole solution begins with the core components that consist of the physical elements that make up the phone such as the digital circuitry, the embedded software, the radio device, and the enclosure packaging (keep in mind this is a simplistic view for discussion's sake).

The whole solution also includes other important elements needed, such as a software application development platform, interface to the wireless communication infrastructure, manufacturing of the product, quality assurance, and customer support for the users of the whole solution. These are the enabling components of the whole solution that are needed to ensure complete customer and user satisfaction when the capability is delivered.

In Chapter 1 we offered a model that shows how program management links execution to strategy by integrating the work flow and deliverables of multiple interdependent projects to develop and deliver an integrated

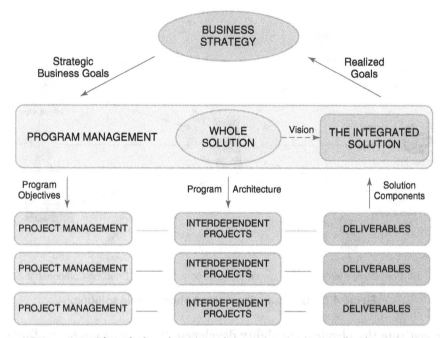

Figure 4.6   The whole solution guides program vision and architecture.

solution. What we didn't discuss at that time was the importance of the whole solution in providing a program vision that focuses the output of the interdependent projects toward a common solution that will enable the achievement of the anticipated business goals. Figure 4.6 illustrates how the whole solution fits within the strategy/execution alignment model.

System design work requires deep collaboration across the functional specialties of a company. By creating the whole solution diagram, a cornerstone element of the program vision is established that provides a visual representation of the integrated solution that will meet the customers' expectations. The whole solution helps to ensure that a holistic view of what it will take to meet those expectations is a driving force in the work of the program team. A strong program vision is in place when each of the members of the program team can see how their work efforts contribute to the creation and delivery of the integrated solution.

THE PROGRAM ARCHITECTURE

The term *architecture* refers to the conceptual structure and logical organization of a system. It includes the elements of the system and the relationships among them.[13] A *program architecture* is therefore the conceptual structure and logical organization of a program. It is composed of the constituent projects of the program as well as the enabling functional components required to create and deliver the whole solution.

To create an effective program architecture, it is helpful to again take a systems approach. In any case where a new capability will be created and introduced into the environment, the program serves as the delivery mechanism for that capability and should be designed and structured in a systematic fashion.

Why is this? If a program is charged with delivering an integrated, whole solution, it has to be architected in a manner in which the whole solution is effectively created. The whole solution will inform one of the primary elements needed of the program architecture. Take for example the simplified whole solution diagram for the creation of a new capability such as a new cell phone, as illustrated in Figure 4.5. The program architecture is easily derived directly from the whole solution, with additional thought toward which of the components needs to be structured as a project, and which components are stand-alone work efforts conducted by members of functional organizations. Figure 4.7 demonstrates an example of how the program architecture might be designed for the cell phone solution.

In this example, all of the core components will be structured as projects, each of which will be led by a dedicated project manager. Additionally, two of the enabling components, the application platform and manufacturing, are efforts that require them to be structured as projects within the program. The other components are not structured as projects, but dedicated personnel from the appropriate functional organizations (or external organizations if the work is out-sourced) must be included as part of the program team.

As we stated in Chapter 1, we like the fact that PMI's program definition includes the following statement: "Programs may include elements of

Figure 4.7 Example of cell phone program architecture.

related work outside of the scope of the discreet projects in the program."[14] If one utilizes the whole solution concept to help define the architecture of a program, the elements of a program that are not managed as a true project, but are essential to meeting the expectations of one's customers, emerge.

This systematic approach to architecting a program works well for strategy-focused and compliance-type programs. But what about for formulated programs? Through our experience, we have found that the process works equally well, but the process for creating the program architecture is not as fluid as it is for the other types of programs.

With formulated programs, one begins with a suite of projects that one wishes to combine into a single program. The *obvious* approach is to create the program architecture from the existing projects. This is not our recommended approach, however. It is best if one starts by identifying the business goals desired, then develop a capability definition that will achieve the business goals (Chapter 6), and then define the whole solution needed to successfully achieve the goals. At that point, the existing projects can be evaluated to determine their relevance to creating the capability and delivering the whole solution, and therefore whether they should be included in the program architecture. Changes to the suite of projects and the charter of each project can then be made as appropriate.

Adding Dimension to the Program Architecture

When utilizing program management to create and deliver the whole solution, it is helpful to view management responsibilities within the program in two dimensions: vertically and horizontally. This concept, as illustrated in Figure 4.8, is a core characteristic of program management. The figure shows a simple example of five components of the cell phone example discussed previously.

Both the vertical and horizontal elements involved in the management of a program are evident in Figure 4.8. First, let's look at the vertical dimension. The work of each function or organization involved in the program is structured as a project and is led by a project manager. These project teams of specialists are responsible for the development and delivery of their respective pieces of the solution. The end result or output of the work accomplished by each project team is known as the project deliverables or outcomes. As an example, the circuit development project team is responsible for delivery of the various circuit boards that make

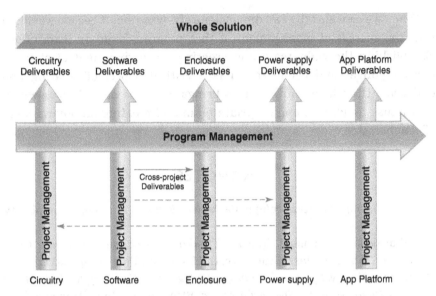

Figure 4.8   Horizontal and vertical dimensions of a program.

up the whole solution; the software project team is responsible for delivery of the software stack (application, drivers, and utilities); and so on for each of the project teams. Again, the project managers are managing vertically, or *within* their specialty domain, to develop their respective element of the whole solution and deliver it to the other members of the program team.

The horizontal dimension of the program is represented by program management, which cuts *across* the project teams and synchronizes and integrates the work flow and outcomes of all constituent projects.

Using systems design to provide an integrated solution prevents the vertical elements of a program from taking a silo approach to developing and delivering their component of the whole solution. Rather, it establishes working relationships *between* the teams of project specialists. To be successful, each project team within the program is highly dependent upon cross-project deliverables from the other project teams. In Figure 4.8, the cross-project deliverables are represented by the horizontal lines between the projects. We call this cross-project interdependency the "space" between the project teams. Management of the space between the projects is the responsibility of the program manager and involves the identification and synchronized delivery of the project interdependencies. This is accomplished through interface definition, cross-project coordination, communication, decision-making, and problem solving.

When we consider the two dimensions of a program, it becomes evident that the successful management of that program requires team effort. Establishing the right team and creating an environment of effective cross-organizational collaboration is a critical aspect of creating and delivering the whole solution. This brings us to the next chapter where we discuss building the appropriate team and creating an environment of collaboration.

ENDNOTES

1. www.nasa.gov/offices/oce/appel/ask/issues/42/42s_challenge_complexity .html.
2. Gharajedaghi, Jamshid. *Systems Thinking: Managing Chaos and Complexity*. Woburn, Mass.: Butterworth-Heinmann Publishing, 2011.
3. Brown, Jimmy. *Systems Thinking Strategy: The New Way to Understand Your Business and Drive Performance*. Bloomington, Ind.: iUniverse Publishing, 2012.
4. http://mib.rbs.com/docs/MIB/Insight/Simplifying-complexity/EIU_report-The_Complexity_Challenge.pdf.
5. Moore, James F., and Jonathan Reese. *Death of Competition: Leadership and Strategy in the Age of Business Ecosystems*. New York: HarperCollins, 1996.
6. Gharajedaghi, *Systems Thinking*.
7. "When Simplicity Is the Solution." *Wall Street Journal*, March 29, 2013.
8. Thoreau, Henry David. *Walden*. Philadelphia: Courage Books, 1990 Reprint.
9. Reed, G. E. "Leaders and Systems Thinking." *Defense AT&L*, May–June 2006, pp. 10–13.
10. Stevens, Richard. *Systems Engineering: Coping with Complexity*. Great Britain: Pearson Education, 1998.
11. Moore, Geoffrey A. *Crossing the Chasm: Marketing and Selling Disruptive Products to Mainstream Customers*. New York: HarperCollins Publishing, 2006.
12. Martinelli, Russ, and Jim Waddell. "Aligning Program Management to Business Strategy." *Project Management World Today*, January–February 2005.
13. New Oxford American Dictionary, Third Edition. New York, NY: Oxford University Press Publishing, 2010.
14. Project Management Institute. *A Guide to Program Management Body of Knowledge*. Newtown Square, Penn.: Project Management Institute, 2004.

Chapter 5

The Integrated Program Team

We have been exploring how systems thinking is a core aspect of program management beginning with the previous chapter. We began with the need to view the output of a program in terms of a whole solution that meets the expectations of one's customers and end users. We then demonstrated how a systems approach to develop a program architecture will facilitate the creation and delivery of the whole solution. In this chapter we continue the systems approach to program management by exploring how to utilize the whole solution and program architecture to create an integrated program team that can effectively execute the program.

An effective program structure is key to realizing the benefits of program management. Without careful consideration of how an organization's programs are structured, many or all of the benefits of program management will be unrealized. Having said this, however, a surprising number of companies have had difficulties implementing an effective and consistent program team structure.

Our research has revealed that one of the most common errors businesses make in implementing program management is failing to understand the difference between a program team structure and a project team structure. Simply stated, projects, especially larger ones, tend to be vertically structured with multiple layers of organization. Programs, by comparison, require system-level coordination, collaboration, and management. Therefore, they need to be flat and horizontally structured to create the cross-project, cross-discipline network necessary to promote effective collaboration, coordination, communication, and decision making.[1] To begin this chapter, we will examine the process involved with forming an integrated program team (IPT).

93

STRUCTURING AN INTEGRATED PROGRAM TEAM

To be successful, a program team must be structured in a manner that facilitates the coordination of its activities and interdependent deliverables, and in a way that promotes effective communication of what is being accomplished by whom and for whom.

As you recall from the discussion in the previous chapter, two key drivers for the formation of the program team structure are the characterization of the whole solution and the design of the program architecture. In effect, the whole solution guides the design of the program architecture and the program architecture guides the structure of the program team. Each of the core and enabling components of the program architecture that will generate deliverables or outcomes required for the program must be represented on the IPT.

The integrated program team consists of three primary entities as illustrated in Figure 5.1: 1) the program manager, 2) the IPT core team, and 3) the extended team.

The particular functions and support organizations that make up the integrated program team are dependent upon the elements and scope of the program architecture. Program team membership may also vary as a program progresses through its life cycle. To remain effective, it is important that the IPT be limited to the key functional representatives within

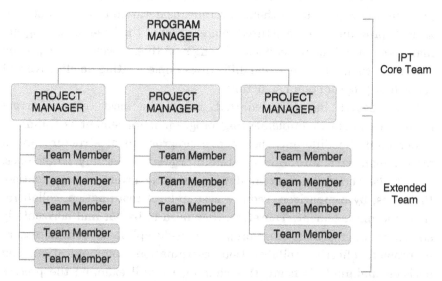

Figure 5.1   The IPT team structure.

an organization. It is recommended that all programs within an organization be structured in a similar manner for consistency.

The IPT Core Team

The IPT core team is the cross-discipline, cross-project leadership and decision-making body of the program that is responsible for ensuring that the program and business objectives, as well as customer satisfaction, are achieved.[2] The IPT core team must become a very cohesive team that has a shared responsibility for the business success of the program. Each member of the core team must be committed to the success of the other members on the team.

The IPT core team consists of the program manager, the project managers, and the functional representatives who provide leadership for the delivery of the elements of their discipline for the program. Collectively the IPT core team constitutes the management team of the program that works under the direction of the program manager.[3]

The size and make-up of the IPT core team is dependent upon the scope and complexity of a program as defined by the program architecture. Typical IPT core team size varies between four to twelve members, including the program manager.

To illustrate, let's look at an example that involves the creation of a product that many of us have come to enjoy using—a tablet mobile device. Figure 5.2 shows a simplified program architecture for the tablet development program.

The program consists of eleven elements: six projects and five program-enabling functions. In the spirit of *form following function*, the IPT core team consists of twelve members: the program manager, six project managers, and five discipline-specific representatives for the program enablers, as illustrated in Figure 5.3.

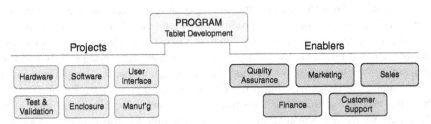

Figure 5.2   Tablet development program architecture.

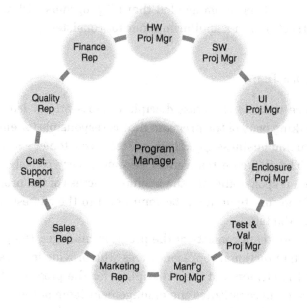

Figure 5.3   Tablet development IPT core team.

This IPT core team is fairly large and is driven by the functional elements needed for the successful development, delivery, and support of the tablet product.

As the leader of the IPT core team, the program manager facilitates the communication and collaboration between the members of the core team (see "An Engineer in Charge of Marketing"). He or she also brokers any conflicts and issues that arise among core team members, between the core team and the functional managers, and between the core team and the stakeholders.[4]

An Engineer in Charge of Marketing

This is a story about Chen Liu, a marketing project manager for a $1 billion company and current member of a product development IPT core team. "I've been with the company for 24 years now. The first 12 years I worked in the laboratories doing technical research, then 8 years as an engineering manager. I moved into marketing four years ago, which was a bit of a tough transition," explained Chen.

He adds, "As the marketing project manager on a core team, I represent the broader marketing function. I coordinate the work of the various marketing functions so we are working in a cohesive manner for the program manager. There's the worldwide MarCom (marketing communications) team, the regional field marketing people, the regional marketing managers, and the application engineers. So my job is to create a marketing project plan, integrate that plan within the master program plan, and coordinate the work of the various marketing people."

When asked about how he works with the other members of the IPT core team, Chen was quick to explain. "We sometimes joke about this in the marketing group, saying we're the center of all activity on a program. But, in reality, all functional teams probably feel the same way because our work is so intertwined. I interface a lot with finance concerning revenue projections and product sales pricing; I interface with manufacturing by providing forecasts that drive materials supply purchasing, as well as identifying the number of demonstration units the field will need; and I work with the customer service team to ensure our customers are fully supported. I also interface a bit with the customer documentation people as a reviewer of the user's manual and with the quality and validation teams to define quality goals based on customer expectations and various regulations. But, the team I interact with the most is the engineering team to make decisions about features and functions, product cost, and ensuring the marketing and engineering requirements are aligned. This is where my engineering background is a real advantage."

Chen concluded by adding, "It's quite possible that I interface directly with every project team represented on the core team."

Chen's story brings to life the realities of how interdependent the project managers and their teams become during the course of a program.

In the IPT structure, the core team members normally report *solid line*, or directly, to their specific department or functional manager, and *dotted line*, or indirectly, to the program manager. Therefore, it is said that core team members wear two hats—the program hat and the functional hat.

This means that the IPT core team members have two sets of responsibilities: program team responsibilities and functional responsibilities. Table 5.1 lists a number of the program and functional responsibilities of the IPT core team members.[5]

The Extended IPT

The extended team consists of individual contributors that make up project teams on the program. They can be assigned to the program on either a full-time or part-time basis dependent upon the work they are to accomplish. They are responsible for ensuring their respective project accomplishes all deliverables to the program within schedule and allocated budget, and with full functionality and performance.

Table 5.1 Core team member program and functional responsibilities.

Program Responsibilities	Functional Responsibilities
Deliver their discipline-specific element(s) for the program	Ensure the functional project team is adequately staffed
Define and execute the requirements of their functional organization	Ensure adequate expertise exists on the functional project team
Develop their project plan and work with the program manager to integrate their project plan into the master program plan	Represent the functional perspective on the program
Manage the work commitments for their respective project team	Ensure project specific objectives are met
Ensure that functional issues impacting the team are raised proactively within the IPT core team	Identify functional risks and manage the risks within their sphere of influence
Work with other IPT core team members to deliver cross-project deliverables	
Assist in resolving problems	
Drive *project* level decisions	
Assist the program manager in balancing the program constraints	
Effectively execute established program processes, methods, and tools	

The project managers lead the extended team and are the primary decision-makers on each project team. It is common for many of the project teams to be organized in the same horizontal manner as the IPT core team.

For example, consider a software project manager who is a member of an IPT core team. In addition to representing the software function on the IPT core team, the software project manager also leads the software project team. The project team is likely to have multiple software-specific disciplines represented, such as operating system, software applications, software drivers, and so forth. Each software-specific discipline is therefore likely to have a team lead that is accountable to the project manager as illustrated in Figure 5.4.

Each team lead will have a set of individuals reporting to him or her for the duration of the team's involvement on the program. The individuals are specialists in their respective functional areas or domains.

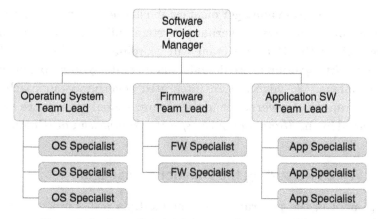

Figure 5.4   Example of software project team structure.

It should be recognized that most members of the extended IPT are execution specialists who are quite happy working in a well-defined environment. It may take significant work to establish the cross-project collaboration needed in the ever-changing environment of a program.

The IPT Program Manager

As adequately represented in the various program management standards and guidelines, the role of the program manager is separate and distinct from that of the project manager. The program manager is the leader of the program team and is held accountable for its success. In this capacity, the program manager is the champion for the program and has primary influence over the work of the personnel involved on the team.

Although there are many aspects to managing a program, there are a few critical aspects that are directly related to establishing a foundation for effective IPT management.

The program manager needs to work with the project managers to establish the methodologies that will be implemented at the project level. A good program manager will then empower the project managers, but hold them accountable to the rules of the methodology they have chosen.

As advocated by guides and standards, program managers should impose a standard reporting process, format, and content to be consistently utilized across the program. The program manager must then set expectations for use of the reporting process to ensure consistent program and project data.

Lastly, the program manager must establish the governance system for the program. An effective governance system will establish the process through which the IPT core team will coordinate and integrate their work, interact together to create value, and create synergy between the team members. We advocate an integrated governance approach that links program-level governance to the organizational governance system as promoted by the *Managing Successful Programmes* guide (Chapter 7).[6]

A Few Words Regarding the Functional Manager

Management of the program occurs at multiple levels of the organization. Executive management sets the business strategy and objectives, the program manager and the IPT core team defines plans and implements the program objectives that help achieve the business objectives, and the project teams are responsible to plan and execute the specialized work to deliver their elements of the program. Additionally, the functional managers ensure that the project teams are adequately staffed and that functional capability is sufficient to achieve the program objectives and realization of the business benefits.

When transitioning from a project orientation to a program orientation, one of the biggest fundamental changes to roles and responsibilities within the firm is that of the organization's functional managers. Under a functional or project-oriented structure, the functional manager has significant power and influence over all or most organized work efforts. However, under a program oriented IPT structure, much of the power and influence for a program shifts to the program manager, with the functional managers providing support to the program for their project team representatives on the program.

Under the IPT core team structure, the functional managers are freed from the daily execution details of a program and can concentrate on the capability growth of their function. They have more time available to focus on their strategic role for the business and their operational responsibilities for hiring, training, and developing the functional specialists within their organization, maintain skills, best practices, and tools to sustain long-term functional expertise. This approach enables the functional managers to make resource and work commitments in support of the program, and assign qualified functional project managers and specialists to represent their function on programs.[7]

Program managers need the help of functional managers to resolve cross-functional issues and conflicts, assist in making key trade-off decisions, keep programs adequately staffed with skilled resources, and provide technical review of cross-project work. For these reasons alone, program managers must establish good working relationships with the functional managers supporting their programs.

STAFFING THE INTEGRATED PROGRAM TEAM

The responsibility for ensuring the IPT is adequately staffed falls upon the program manager. He or she is directly responsible for ensuring the IPT core team is fully staffed, while the project managers are directly responsible for staffing their respective project teams. This is not to say that the program manager does not assist the project managers in obtaining the qualified personnel, if necessary.

The program manager is not responsible, however, for selecting the resources for the entire program. Rather, he or she ensures each project manager has the ability to staff their project team with the number of resources required, as well as fulfill critical skill requirements. Where gaps exist in these areas, the program manager will work with functional management, executive management, and the human resources department to fill resource gaps. If critical gaps cannot be filled, changes to the program plan may be required. The program manager should consistently ask for project team staffing reports to ensure the IPT is adequately staffed, especially if there is a change in personnel or a change in requirements and scope.

CRITICAL FACTORS FOR IPT SUCCESS

One of the primary characteristics of a program is that it is highly integrated and synergistic in nature and is truly a case in which the sum of the parts is more valuable than any of the parts on its own. Additionally, programs tend to be dynamic, with intense cross-discipline integration in which the actions of one project team affect, support, and reinforce the other project teams involved with the program.

Like all other teams, program teams must be effective in team communication and cross-team coordination for repeatable success.

Team Communication

Highly effective teams communicate clearly, consistently, and frequently within the team structure, as well as with stakeholders outside of the team.[8] Effective communication permits the program team to make better decisions, evaluate information between the project teams to assess impacts on interdependencies, deal with inter-team conflicts, and keep key stakeholders apprised of program status. The program team structure must enable open, nonhierarchical, and clear communication channels within the team. The program manager and the team leaders must be adept at both vertical and horizontal communication. From a horizontal perspective, the program team must be able to effectively communicate across a wide spectrum of groups and disciplines, including, in some cases, outside suppliers and partners of the firm (see "The Russians Join Us Late at Night"). From a vertical perspective, the program team must be able to communicate effectively with senior management at the top of the organization down through the organization structure to the individual contributors on a program, such as manufacturing operators on the production floor.

The Russians Join Us Late at Night

"Communication is the key," said Sri Rastogi. Rastogi is a project manager on a program that is geographically dispersed, with part of the team in Portland, Oregon, part in Houston, Texas, and part in Moscow, Russia. Rastogi knows the importance of communication in geographically dispersed teams. He also knows changes are sometimes needed to create opportunities for better communication.

As he explained, "There is an eleven hour time difference between Portland and Moscow. So finding a good time to communicate in person is tough. One of the things the Moscow team has done is shift their work day. They now come in about 10 or 11 o'clock in the morning, then go home anywhere between 8 and 10 o'clock in the evening. We now have overlap at the end of the day where we can usually find people in the office."

Instant messaging technologies have also helped a lot. "I log in from home for an hour each night and turn on my instant messenger. If anyone in Russia needs to contact me during their morning, they can do so, and I'll respond immediately. I don't know how it is for the rest of the company, but the fact that I make myself available at 11 o'clock at night on a daily basis is a necessity to help communication channels stay open on a geo-dispersed team. It's not rocket science, but it works!"

Cross-Team Coordination

An effective program team must be able to coordinate and integrate many complex activities and deliverables. Traditional hierarchical and functional team structures are inadequate to address the high level of cross-project coordination that needs to take place in a timely and cost-effective manner.[9] The program team structure must horizontally integrate the project teams to enable the program manager to effectively coordinate and channel the work of the teams toward a common output.

Figure 5.5 illustrates the triangulation of collaboration that takes place on an IPT core team. Directions, decisions, and cross-team issue brokering come from the program manager to the project managers; cross-project communication and work coordination occurs between the project managers; and status, decision consultation, and issue escalation come from the project managers to the program manager.

Figure 5.5   IPT Core team collaboration triangulation.

As illustrated in Figure 5.6, when the third element of the IPT, the extended team, is added, a more traditional project management structure becomes evident at the lower level of the program pyramid.

Decisions, direction, and program pass-downs come from the project managers to the extended program team members. Work status, issues,

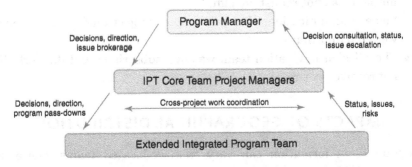

Figure 5.6   Blended vertical and horizontal coordination.

and detailed risks come from the extended project team members to the respective project managers. The large majority of the cross-project coordination occurs at the core team level, even though some detailed interaction will occur between individual contributors on the project teams.

The horizontal, cross-project structure that the IPT approach provides is key to establishing the shared responsibility and understanding needed to achieve program success. While specialists within an organization's functions are used to collaborating on a daily basis, on a program, coordination between the functional specialists is also necessary.

Let us look at an extreme example of the importance of good cross-project communication and coordination. Nine months after being launched, the Mars Climate Orbiter was lost during its first pass around the red planet. The spacecraft became the victim of poor communication between project teams on the program. One team programmed navigational data to be sent to the Orbiter in English units (feet, inches, and pounds), while another team programmed the Orbiter to receive and interpret the data in metric units (meters, centimeters, and kilograms). The miscommunication resulted in the Orbiter entering the Martian atmosphere at 57 km instead of the intended 140–150 km, and the $125 million spacecraft was destroyed by heat caused by atmospheric friction. Among the primary contributing factors for the loss were the following, which reinforce the importance of good cross-project coordination and communication:[10]

- The operational navigational team was not fully informed about the details of the way that the Mars Climate Orbiter was pointed in space.
- The systems engineering function within the program that is supposed to track and double-check all interconnected aspects of the mission was not robust enough.
- Some communications channels among project engineering groups were too informal.
- The mission navigation team was oversubscribed and its work did not receive peer review.

IMPACTS OF GEOGRAPHICAL DISTRIBUTION

Program team formation can vary from co-located teams where all members of the program team reside at the same site, to geographically dispersed program teams where members are participating from various physical locations across a nation or around the globe.

A *geographically distributed team*, also known as a *virtual team*, is comprised of team members who participate on the IPT, but are separated by time, distance, and sometimes cultural differences. The bond that holds the team members together is the common purpose, vision, and leadership of the program manager. No doubt, the challenges and barriers to leading a geographically distributed team are significantly greater than leading a co-located team due to several factors that include managing across time zones, communication challenges, language barriers, and cultural considerations.

Participation, Collaboration, and Integration

It is important to note, however, that from the program manager's point of view, the foundational elements of effective team leadership and management will continue to work well and are applicable across the entire spectrum of team composition options. Fortunately, success begins with proficiency in the basic principles of team leadership along with the knowledge of how to extend these basic principles into a more complex distributed team environment with inherent differences in team dynamics and communication needs.

Some of the more important leadership principles that must be extended to the distributed team include the following:

Creating a Common Purpose

A geographically distributed team must have a common purpose as the basis for collaboration and participation. A clearly defined common purpose is instrumental in removing ambiguity and should answer four questions:

1. What is the purpose? (The business benefits)
2. What does the end state look like? (The whole solution)
3. What do we need to accomplish to get to the end state? (The business success criteria)
4. How will success be measured? (The critical success factors)

Establishing a clear program vision is critical to creating the common purpose needed.

Establishing Team Chemistry

Establishing team chemistry is considerably more complicated within a geographically distributed team because of the cultural diversity considerations and communication challenges caused by language and time zone differences. There are a number of things that successful distributed teams do to accelerate team cohesion, including establishment of team norms for how the team will interact and operate. Additionally, successful teams generate as much social presence and face-to-face interaction among team members as the organization will allow. A third condition to establish team chemistry is the utilization of team-based electronic tools to enhance communication and collaboration.

Building and Sustaining Trust

Trust within a team is the foundation for effective collaboration. For a team to reach peak performance, considerable attention must be paid to building trust among team members and between the program manager and the team members. Trust in the program manager is critical. At a minimum, a program manager must perform competently, follow through on commitments, display concern for the well-being of others on the team, and behave in a consistent manner.

Empowering the Team

Empowering key members of a virtual team to operate autonomously within the boundaries set by the program manager is more critical than with co-located teams because of the physical isolation and distance from the rest of the team. Bottom line, many times some of these team leaders and individual contributors to the program are on their own at their specific company site separated from the rest of the program team that may be hundreds or thousands of miles away.

Team empowerment for a virtual team means giving the project managers and program team members the responsibility and authority to make decisions at the local level. For the program manager, however, greater empowerment of geographically dispersed team members means greater risk for obvious reasons. The greater the empowerment granted to the distributed team members, the less control the program manager maintains.

The most powerful tools available for a program manager to use in establishing effective team empowerment are clearly defined deliverables, identified owners for each deliverable, and decision boundaries based upon success criteria associated with each deliverable.

Communicating Effectively

The three primary elements of effective communication are: 1) listening, 2) reporting, and 3) facilitating. In a geographically distributed team, the selection of the communication tools such as teleconferencing, email, websites, and blogs must be balanced to optimize communication in order to drive the program team's participation, collaboration, and required integration of the team's efforts.

Culturally Blended Teams

Another key difference between co-located and geographically distributed teams is that typically many distributed teams span multiple national, cultural, and linguistic boundaries, contributing additional challenges to the program manager. It therefore becomes necessary for the program manager to learn and embrace the cultural norms, beliefs, and behaviors associated with each country that is represented on the program team. The program manager in this environment must be the champion for cross-cultural leadership and become a role model for good culturally sensitive behavior (see "My Job Was to Integrate Two Cultures").

My Job Was to Integrate Two Cultures

"All my professional life I have dealt with software development—banging out code," began Jerry Dorsey, now one of several project managers for the geographically and culturally dispersed Dacia program. "I have always managed local, co-located software development teams. So I was stunned when my boss summoned me and asked me to manage a program with an out-sourced development team in Romania. I think I asked the same question three times—Romania?" Jerry had expressed interest in moving into a program management role, so his boss worked with the program office director to move him into a role that required integration skills—in this case, solving a cross-culture problem by integrating the Romanian team with the U.S. team.

Jerry continued, "At first I was shocked, since I didn't know the first thing about cross-cultural integration, but I began communicating with members of the Romanian team to learn about how they worked and what they valued." Jerry soon discovered that corporate culture and national culture often collide. "The Romanians were used to being tasked," said Jerry. "They had an attitude toward me that 'he's the boss,' and, therefore, I should have all the answers. The concept of brainstorming solutions, which is a common part of our company culture, was completely unknown to them. Because they lived under communism dictatorship for so long, they were used to people telling them what to do and just doing it."

Jerry continued, "They will also never say no. I could just give them more and more to do and they'd try to get it all completed. So, I had to learn how much I could actually give them by monitoring the progress of their deliverables. As long as they met their deliverables on time, I figured they weren't being overtasked."

The biggest lesson for Jerry, however, had little to do with managing the development of the software. As he explained, "Building strong personal relationships was the most critical element in integrating the Romanian team into our company and program culture. We were able to bring the key technical leaders from Romania to the United States early in the planning phase to meet and interface directly with their United States counterparts. There's no better way to build mutual trust! I also made a point to travel to Romania once every two to three months to get to know the Romanians and make myself directly accessible to them."

"At the end of the day," concluded Jerry, "this was a great experience for me personally and for my career. I got first-hand experience on what it means to be a program manager, and it's definitely an avenue I'd like to continue to pursue."

At a minimum, the following set of guidelines will help to establish a culturally balanced team environment, which recognizes and utilizes cultural differences as a strength:[11]

- Suspend judgments. Do not make generalized or stereotypical assumptions about team member's cultures.
- Admit that we don't know. Much of our current knowledge regarding other cultures may be incorrect.
- Show empathy. By listening and caring about others, we learn how other people would like to be treated.
- Systematically check our assumptions. Ask for feedback and constantly make sure you clearly understand the situation.
- Become compatible with ambiguity. Accept the fact that globally dispersed teams will be more complex and that many things will not be totally clear.
- Celebrate diversity. Recognize and espouse the value of differing viewpoints, opinions, and ways of doing things.

With an integrated program team structure in place, an organization becomes well positioned to achieve the business benefits that program management offers an organization such as rapid time-to-benefits, effective management of complexity, establishment of consistent processes and tools for cross-team collaboration, and effective management of resources across the program. Equally important, a program manager has an effective team structure that forms the basis for successful management of a program, the subject of the next chapter.

ENDNOTES

1. Martinelli, Russ, and Jim Waddell. "Demystifying Program Management: Linking Business Strategy to Product Development." *PDMA Visions*, January 2004, pp. 20–23.
2. Martinelli, Russ, and Jim Waddell. "Program Management: Linking Business Strategy to Product and IT Development." *Project Management World Today*, September–October 2003.
3. McGrath, Michael E., Michael T., Anthony, and Amram R. Shapiro. *Product Development: Success through Product and Cycle-time Excellence*. Stoneham, Mass.: Butterworth-Heinemann Publishers, 1992, p. 206.
4. Ibid., p. 84.
5. Ibid., pp. 87–90.
6. Office of Government Commerce. *Managing Successful Programmes*. 3d ed. Norwich, UK: Office of Government Commerce, 2007.
7. Martinelli, Russ, and Jim Waddell. "Program Management, Part II." *Management RoundTable Best Practices Report*, November 2010.
8. Eikenberry, Karen. Elements of a High Performance Team, Sideroads website: www.sideroad.com/Team_Building/high-performance-teams.html.
9. Mohrman, Susan A., Susan G. Cohen, and Allan M. Mohrman, Jr. *Designing Team-Based Organizations: New Forms for Knowledge Work*. San Francisco: Jossey-Bass Publishers, 1995, pp. 5–6.
10. NASA website: http://nssdc.gsfc.nasa.gov/database/MasterCatalog?sc= 1998–073A.
11. Martinelli, Russ, Tim Rahschulte, and Jim Waddell. *Leading Global Project Teams: The New Leadership Challenge*. Ontario, Canada: Multi-Media Publications, 2010.

Chapter 6

Managing the Program

It is well established that program management's greatest contribution to an organization is the delivery of business results. Managing a program, therefore, involves coordinating and integrating the work of others to create the whole solution that, when delivered, becomes the means to capture business value. As we have learned, however, this is many times a difficult undertaking due to the ambiguity and the level of uncertainty associated with the environment in which a program exists. To effectively manage a program with high ambiguity and uncertainty, structure is needed to guide a program's journey from strategy development to benefits delivery. Caution must be exercised, however, to establish the *appropriate* level of structure so a program manager's ability to navigate the amount of change that is normally encountered during the life of a program is not constrained.

Instead of thinking in terms of process for establishing program structure, we recommend thinking in terms of establishing a structured *framework*. Through our work with numerous companies and organizations, we have witnessed the advantages of using a framework to provide the high-level guidance necessary for managing one's programs. But what type of framework is needed? To answer this question, you have to once again look at the primary outcome from a program—delivering business benefits. As a program moves from inception to end-of-life, it is guided by a series of critical decisions. These decisions have both strategic and operational implications, but the intent of each decision is to evaluate the viability of a program to deliver the intended strategic business benefits and decide on a course of action for the next program cycle. Business decision management, therefore, is at the heart of managing a program and is done so in context of an organization's program life cycle. We have found that, in practice, a well-defined decision framework is valuable in providing the high-level structure needed to effectively manage one's program.

In this chapter, we provide an example of a business decision framework and discuss the critical aspects of managing a program within that framework.

A BUSINESS DECISION FRAMEWORK

During the course of a program, a program manager will have to contend with unexpected changes in the market and organizational environment, a large number of uncertainties and assumptions that have to be tested and vetted, and multiple influential stakeholders with opposing views.

For these reasons, a robust business decision framework provides the flexibility necessary to enable an adaptive management process that allows for changes in the program as new information comes in, and at the same time provides anchors to align stakeholders on the critical business decisions necessary to successfully manage a program.

Figure 6.1 illustrates an example of a program-level decision framework that is based upon the critical business decisions associated with a program.

As a program progresses through time, the program manager leads his or her program team through a series of work cycles. Each work cycle is unique in intent and scope, and is focused on completing the work necessary to make informed decisions at each of the major business decision checkpoints. Each organization is unique, with its own set of business practices, and therefore, each organization will have its own version of a business decision framework. However, care must be taken to define a decision framework that is coherent in that it integrates the inputs and work of the various business functions within the organization.

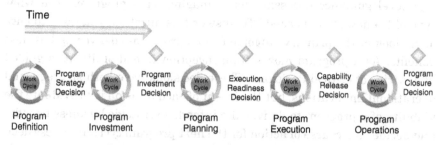

Figure 6.1 Business decision framework.

Many organizations have a formal life cycle that they use to guide how a program progresses from ideation to closure. This is good, but what we consistently see is the bulk of the effort and process definition lies in defining the various work flows and outcomes during each stage of the process, and very little forethought and effort dedicated to the business decisions which need to be made at the culmination of each stage.

What if the way we think of our lifecycle models is reversed? Instead of viewing them as processes consisting of stages of work with decision 'gates' intended to allow work effort to transition from one stage to the next, we view them as a series of business decisions with iterative work cycles designed to prepare for and successfully transverse each business decision as it is encountered. The business decision framework presented in this chapter can be utilized in this manner, and in practice is flexible enough to be adaptable to a waterfall process, agile process, or any process in between.

Enabling Effective Decisions

As we explored in Chapter 3, one of the key benefits of program management is that it helps to align execution output with business strategy. The business decision framework is a program manager's guiding structure to ensure that this alignment is accomplished. As illustrated in Figure 6.1, as a program progresses through time, the decisions are initially strategic in nature, then become increasingly more operationally focused.

The framework alone is insufficient, however. One must be aware that each decision has two critical elements: 1) making a decision, and 2) implementing a decision.

Decision making involves understanding the problem, opportunity, or job to be completed, the criteria that will be used to make the decision, the decision options available, and choosing a course of action. Implementing a decision involves holistically planning the work necessary to carry out the decision, assessing progress toward completing the work, and ensuring work output culminates in an outcome that supports the decision.

It's Not about the Project Methodology

We have listened to the ongoing debate that has been occurring in the project management community over the virtues of iterative (agile) versus linear (waterfall) methodologies for managing a project. Fortunately, we have not witnessed the same debate within the program management community as programs are *both* linear (time is linear) and iterative (work and learning are iterative) in nature. This truism is demonstrated in Figure 6.1, which shows the major program decisions occur over time, but the work that occurs between decisions is normally iterative.

If a business decision framework is utilized as the guiding framework for a program, the methodologies employed at the *project* level are not a significant factor. What *is* significant at the *program* level is that the work output from each of the projects within a program must synchronize and integrate with the outputs from the other projects—regardless of the project methodology used to generate the outcomes. For example, even though a software project team on a program employs an agile project management methodology and a hardware project team on the same program employs a waterfall methodology, the outputs must synchronize over time and integrate into a holistic solution at the program level ahead of the critical decision checkpoints.[1]

This type of approach requires strong collaboration between the program team and the firm's management team so there is both business management and program management involvement in the realization of business benefits. The business decision framework should be used to guide this collaborative arrangement. Referring to Figure 6.1, let's look at each of the key business decisions in more detail.

DECISION CHECKPOINT: PROGRAM STRATEGY

The first major business decision on a program is normally strategic and conceptual in nature. It centers on ensuring that both the program strategy and the capability to be generated by the program align to the strategic business goals of the organization. Due diligence at this business decision point is critically important to ensure that the organizational strategy and goals are in alignment with the execution of the program. It is common to hear stories and lessons learned of frustration about misalignment between a program's output and the business goals intended; the root cause usually points to the lack of documented proof of alignment to business goals.

The cycle of work geared toward preparation for the program strategy decision is commonly referred to as *program definition*. Program definition is an extension of the strategic management process and is critical in coalescing an organization's business functions and influential stakeholders to a common strategic vision and program strategy.

Program definition is a learning and decision-making process that is critically important, but disproportionately difficult due to the lack of quantifiable data. This stage of a program is commonly referred to as the *fuzzy front end* because people are working with ambiguous information and assumptions to formulate concepts and to predict multiple scenarios that may play out anywhere from one to ten years in the future. As a marketing research analyst from Microsoft Corporation told us:

> When I ask our customers what they envision a personal computer or tablet being able to do in the next five years, they look at me somewhat strangely and tell me that they're not sure. Right now they're just trying to focus on staying in business and staying relevant in their business.

Regardless of the well-documented challenges associated with the fuzzy front end, some of the greatest opportunities for business success lie in the work and the output of this initial part of a program.

In many organizations, the program definition cycle is funded as a separate activity from the remainder of the program. The intent is to gain funding approval for the next cycle of work by aligning the program with the strategic business goals of the organization and generating concepts for the program output (product, service, infrastructure capability, or change transition end state).

Managing the Program Definition Cycle

One of the advantages that a program-oriented organization possesses is that they have invested in the development of strong program managers who can participate and add value in the earliest parts of a program. This advantage enables programs to be better aligned to business strategy and move more effectively through the program definition learning cycles. The program manager will need to facilitate the collaboration of a cross-discipline team in order to create the program strategy and capability concept. The program definition team tends to be small (three to six members on average), but their specialty expertise, jargon, and professional perspective of what they view as important, and what is not, will vary considerably. Having a strong, independently minded program manager with multi-discipline experience and perspective as a leader is an advantage.

During program definition, the program manager's responsibility is to ensure that the work outcomes necessary to support a high-quality business decision are complete and that the critical stakeholders are aligned to the program strategy. To accomplish this, the program manager will need to lead the program definition team through a number of work activities (described below) that culminate in the preliminary business case.

Defining the Program Strategy

The program strategy describes how the program team will achieve the strategic business goals desired from the outcome of the program. It is established by defining the program objectives and the capability output (product, service, infrastructure, or change transition), that once achieved, becomes the end state for program success.

As discussed in Chapter 3, the program objectives establish alignment of a program to the business strategies of an organization, and provide the guiding principles that help the program manager and other key decision makers understand the scope of the program, make key trade-off decisions, and determine resource requirements. To be effective, objectives should be specific, measurable, realistic, and time related.[2]

Defining the New Capability Concept

To achieve the program objectives, the program output should result in a new capability that becomes the *means* for achieving the desired outcomes. Capabilities normally come in the form of a new product or service to be offered to a firm's customers, a new internal capability that will positively affect a firm's effectiveness and efficiency, or a new change initiative that will create positive transformation either internally or externally to the organization.

Deriving a capability concept, or set of concepts, is an iterative process that involves the integration and synthesis of the ideas from the subject matter experts on the program definition team. When leading the ideation process, the program manager should focus the output toward the program objectives identified. This involves the following: facilitating the cross-discipline collaboration, setting periodic concept reviews in front of primary program stakeholders to create tension in the system and expedite the concept definition, continually collecting and communicating any changes to the initial program objectives, and testing the concept feasibility from a business perspective.

Developing the Preliminary Business Case

A fully complete business case is not always necessary for the first cycle of work on a program, and in fact may be wasted effort if the program objectives are not agreed upon by influential stakeholders. We recommend a separate business decision fully dedicated to the approval of the program business case following approval of the program strategy. That being said, a preliminary business case is needed to support the program strategy decision. Critical elements include information on the potential benefits, business success factors, costs, and potential risks associated with the program being proposed.

Benefits Identification. Early benefits identification is an iterative learning process and may consume a disproportionate amount of the program definition team's time and effort. It begins with understanding the strategic business goals driving the need for the program, and then dissecting the goals with the input from the program stakeholders, customers and end users, strategic suppliers, and others. The result of this effort is what PMI refers to as the Benefits Realization Plan.[3]

Business Success Factors. In Chapter 2 we described how we delineate business benefits into business *value* and business *results*. Establishing the right measures for delivery of the business results is a key activity of program definition. We firmly believe in the old adage that says, "What gets measured, gets done."

The business success factors transform the business results derived from strategic goals into a specific statement of success that guides the execution engine of the organization as to how to plan and execute their work. We define business success factors as *the set of quantifiable measures that describe the successful achievement of a program's business results*. It is the business success factors that bind the activities of strategy setting to those of strategy execution.

In Chapter 3, we discussed the translation from strategic goals to program objectives that needs to occur in order to effectively align a program to the business strategy. Business benefit identification is at the heart of that translation effort. The translation goes as follows: Strategic business goals to business results and from business results to program objectives. Table 6.1 provides an example of this translation process.

Once a translation of goals to benefits to objectives is complete, the program definition team should validate the findings by working

Table 6.1   Strategic goals to business results to program objectives.

Strategic Goals	Business Results	Program Objectives
Be first to market with the new web-based service	1. Greater than 50% market share 2. 47% profit margin 3. $5 million in revenue during the first full year	1. July 2017 market introduction date 2. 85% customer satisfaction rating during user trials 3. Available in 5 target languages

backward to ensure that the program objectives will indeed result in the desired business results; then, that the set of business results identified will achieve the strategic goal. When the team is satisfied with the quality of the translation, they can then move to the next strategic goal and translation process.

Cost Evaluation. With the initial benefits of the program established, the preliminary business case should also estimate the cost to execute the program. Program costs represent the investment that the organization must make in the program.

At this stage, a high-level estimate of program funding, capital expenditures, and resources required should be completed. Additionally, if third party organizations are involved—such as strategic partners—an estimate of their investment cost also needs to be provided.

With both the initial benefits and costs identified, the preliminary business case can contain a high-level estimate of program cost versus benefits.

Risk Identification. To promote a risk-based decision, the preliminary business case should include an analysis of the significant risk events that may negatively impact the achievement of the business results. The objective of this exercise is to identify as many unknown events that may occur, and to assess them from the standpoint of impact to the stated business success factors.

A full risk management plan is not necessary at this early juncture of a program. Rather, the intent here is to proactively identify the critical risks that the definition team feels will likely have a negative impact on the program strategy, and to bring this knowledge into the program strategy decision.

Aligning Stakeholders

Program definition provides the first opportunity to identify the various stakeholders of the program, evaluate who is critical to the success of the program, and analyze their needs. A stakeholder is defined as anyone who has a vested interest in a program.[4] More importantly for the program manager, a *critical* stakeholder is *anyone who can influence, either positively or negatively, the outcome of the program.*

Stakeholders play a key role in the program definition process by helping to identify the business benefits expected, providing direction on the capability concepts, and defining the program objectives. The value of early stakeholder identification normally becomes apparent to the program manager. Not only does one get a sense of stakeholder expectation for the program, but also of the opinions, perceptions, and personal agendas that may affect the outcome of program definition and possibly the outcome of the program as a whole.

Proficient program managers never leave the definition of program success up for debate until the end of the program. Instead he or she moves the debate on what constitutes success to the first decision point of the program. It is rare that early consensus is reached among the influential stakeholders on how to measure program success—in fact, consensus may never be reached. The program manager is charged with brokering the negotiation among influential stakeholders on how to define program success through the business success factors.

Some advocate the use of workshops to engage all influential stakeholders in a single session to establish the business success factors. In practice, this becomes difficult for a couple of reasons. First, executives are busy, and trying to align schedules for a two- to four-hour workshop can be an impossible task. Second, not all program managers possess the high level of credibility and skills needed to lead an executive-level negotiation workshop. Rather, we have found it most effective for a program manager to work with the stakeholders individually and to leverage the influential program champions in establishing the business success factors prior to the program strategy decision.

The Program Strategy Decision

Having completed the outcomes required during program definition—program objectives, capability concepts, preliminary business case, and stakeholder alignment,—the team is prepared for the program strategy decision. In strong program-oriented organizations, the program manager will lead the presentation of the program strategy and preliminary business case to the organization's senior management. In many cases, he or she will have other members of the program definition team present the detailed information for which they are the specialist—such as the financial analysis and capability architecture concepts.

The outcome of the program strategy decision can play out in a number of ways. Three primary options are available to the executive decision maker:

1. Do not approve the program strategy or the capability concept with direction to redefine the strategy or concept.
2. Do not approve the program strategy or the capability concept and terminate the program.
3. Approve the program strategy and the capability concept and move into the next cycle of work. Approval may come with sponsor and stakeholder redirection.

If approval is achieved, the program team is given the authorization and funding to develop the analysis and material to support the full program business case in preparation for the program investment decision.

DECISION CHECKPOINT: PROGRAM INVESTMENT

Following a decision on the program strategy and one or more capability concepts addressing how the program strategy will be achieved, the next major business decision on a program is normally the program investment decision.

The intent of this business decision is to evaluate whether there is sufficient justification to approve funding for a program. If a decision is reached to fund a program at this point, the program will normally be added to a firm's implementation portfolio (many times called an R&D portfolio) where it will likely be evaluated against other programs for allocation of resources.

The cycle of work ahead of the program investment decision is focused on selecting a single capability concept and on developing the detailed business case for the program.

Managing the Program Investment Cycle

The program investment cycle can be characterized as one of hypothesis setting, rapid learning cycles, and validating outcomes to close on a single capability concept and business case that demonstrates the ability to realize the intended business benefits.

At the end of this work cycle, the program manager is responsible for ensuring the work output from the team supports the requirements for a high-quality business decision.

Defining the Final Capability Concept

As in the program definition cycle, the program manager is critical in facilitating the work of a cross-discipline team charged with coming to closure on a final capability concept. Even if a single concept was established prior to the program strategy decision, additional work will likely be required to validate the assumptions surrounding the proposed capability.

The process to accomplish this is best described as an iterative learning process where one seeks to understand how well a capability might satisfy customer, user, or organizational needs, how well it supports the program objectives and provides the means to achieve the business benefits, and how difficult it will be to create and implement.

The challenge for the program manager is to lead the team to convergence of the capability concept as rapidly as possible, know when to prevent additional learning cycles from occurring, and document the remaining assumptions for the next major cycle of work. Periodic reviews with the program sponsor and key stakeholders can be helpful in overcoming this challenge and maintaining a sense of urgency. Additionally, ongoing involvement and input from customers and users of the new capability is necessary.

Documenting the High-Level Requirements

As part of the final concept definition process the high-level requirements for the capability are identified and documented. The importance of diligently defining requirements cannot be overstated, as requirements are the basis for program planning and capability development activities. Poor requirements will result in poor planning. In turn, poor planning will result in poor execution and rework, which may result in failure to realize the business benefits of the program.[5]

The program manager is responsible for collecting and documenting the high-level requirements during this phase of the program, and should work with the content experts who represent the various disciplines on the program definition team to collect the requirements. To support the program investment decision, at a minimum, the high-level requirements must be documented in sufficient detail to describe the capability concept, key features and functions, end user and customer needs, and cost and financial targets as applicable.

Additionally, the high-level business requirements—the business success factors—that were first identified in the previous cycle of work should be further refined and included in the program business case.

Developing the Detailed Business Case

As the business manager on a program, the program manager leads the development and documentation of the detailed program business case. He or she must ensure that the business case contains the appropriate level of detail to support a funding decision, and that it is presented in the language of executives—the language of business.

The purpose of the detailed program business case is to demonstrate that a program and its capability will support the strategic goals and anticipated business results for the enterprise.

It is a guiding document used by a firm's program sponsor and other senior managers to assess the feasibility of a program and to repeatedly assess the program's path to success. For the program team, the business case is its primary chartering document that defines a successful end state.

As the advocate for the firm's business needs on a program, the program manager should lead the effort to develop and defend the detailed program business case. At a minimum, a high-quality program business case should include the elements shown in Table 6.2.

On many programs, the program business case is developed at a time when more is unknown about the program than is known and has been validated. For this reason, the program business case should be treated as a living document that is periodically reviewed, tested, and updated throughout the life of a program.

Stakeholder Engagement

Since multiple stakeholders will likely have a voice in whether or not a program is funded, stakeholder management is a critical role for the

Table 6.2   Minimum elements of a program business case.

Business Case Element	Description
Program purpose	A succinct statement of the business benefits driving the need for the investment in the program
Value proposition	A succinct statement characterizing the value to be delivered (quantified when possible)
Business success factors	The set of quantifiable measures that describe business success for the program
Detailed cost analysis	The investment cost of the program
Cost/Benefits Analysis	Evaluation of the program return on investment
Critical assumptions	The events and circumstances that are expected to occur for successful realization of the program objectives
Program timeline	Critical program milestones and timing expectations on the part of key stakeholders
Risk analysis	A thorough analysis of the risks that may prevent realization of the business benefits of the program

program manager to fulfill. The existence of differing needs, desires, and competing agendas between key stakeholders may never be as high on a program as when the major investment decision is approaching. A program manager should expect to spend significant time engaging with his or her stakeholders.

Stakeholder engagement will involve gaining agreement on the program value proposition, gaining commitment of resources to execute the program, and addressing the concerns of the stakeholders as best possible. Program champions can provide a wealth of support in executing your stakeholder strategy at this point in the program—experienced program managers never forget to engage them.

Establishing Program Governance

Most companies systematically monitor the achievement of their investment in programs and other work efforts in order to increase the probability of successfully achieving the expected returns (or value) from these investments through the use of their internal systems, policies, and guidelines that govern these activities. Some companies formalize these procedures into comprehensive governance policies and benefits management systems to provide the structure needed to properly oversee these activities. In some corporations, this may be managed through a governance board or steering committee. Other companies may assign this responsibility to one or more of their executives.

Well designed and implemented governance procedures demonstrate to executives and program stakeholders that their investment in the program will have appropriate oversight and control. A robust program governance system has three primary functions:

1. Establishing and maintaining the program strategy based upon the strategic business goals.
2. Ensuring the right structures are in place to achieve the program strategy.
3. Monitoring and directing the program to make sure the stated program objectives and business benefits are realized.

There is a distinct difference between program and project governance processes due to the nature of programs. Governance of programs needs to focus on both strategic as well as operational progress and results, while project governance primarily focuses on control to ensure execution of scope, time, and budget constraints.[6]

Program governance must also oversee the management authorization and support for dynamic change and impact to strategic business and operational objectives of the program as a result of shifts in the market and business strategy.

The Program Investment Decision

Having completed the outcomes required during the program investment process, the team is prepared for the program investment decision. In program-oriented organizations, the program manager will be front and center in presenting the capability concept and the detailed program business case to senior executives and other key stakeholders. However, it is good practice to also include critical team members who were involved in the program investment process, especially if critical questions are presented for which the appropriate discipline specialist is better qualified to answer.

Normally, three decision options are available to the executive decision-maker(s):

1. Do not approve program investment with direction to the program team to further refine the capability concept or the program business case.

2. Do not approve the program investment and terminate all further work effort on the program.

3. Approve the program investment with direction to allow the program to progress to the next primary cycle of work. Approval may come with sponsor and stakeholder redirection.

If approval is achieved, the program team is given authorization and commitment of funds and resources to develop the integrated program plan in preparation for the execution readiness decision.

DECISION CHECKPOINT: EXECUTION READINESS

To this point in a program, the program team has created a strategy to achieve a set of strategic business goals, and has developed a sound business case justifying the investment of a firm's resources. In essence, the program manager and his or her team have been operating as part of the business engine of the firm as described in Chapter 3. The next major business decision on a program—the execution readiness decision—centers on ensuring that the program team has a plan for execution of the program strategy.

"Planning is everything." That's how one program manager from Ford Motor Company described the importance of program planning. However, it seems to be innately attractive to many individuals to just jump in, start taking action, and doing things. Action-oriented individuals often have a difficult time with planning activities. How many times have you heard it said, "Let's just do it; we know what needs to be done and how to do it." Many believe that spending time doing detailed planning is a waste of time.

However, most seasoned program managers have learned through experience that early and effective planning prevents poor performance and problems such as quality issues, missed commitments to customers, and delayed realization of business benefits.

Program planning is about laying the foundation for the execution of the program and coupling execution to the business strategy of an enterprise. Through development of the integrated plan, the program team will demonstrate how the capability concept will be created and how the business benefits described in the program business case will be realized. This information is crucial for an effective execution readiness business decision.

Managing the Program Planning Cycle

The program planning cycle involves the work necessary to structure and organize the program to create and deliver the whole solution, and establishes an integrated, cross-project plan to achieve the business benefits anticipated. Initial program planning happens at a time when data is at a minimum and assumptions are at a maximum. Planning therefore should be approached as both a learning and iterative process. At the end of the work cycle, the program manager is responsible for ensuring the right team and right plan are in place to support a high-quality business decision.

Structuring the Program

The program planning process begins with establishing structure and organization. First, the program architecture consisting of the projects and support functions for the program has to be defined. Next, the IPT core team has to be established and staffed with the appropriate team members as described in Chapter 5. Finally, the program scope, which defines the deliverables and outcomes for each of the projects and support functions, has to be developed.

Before these activities begin, however, the program manager should ensure that all the information about the program that was created from the earlier work cycles is carried forward and shared with the program planning team. A clear understanding of the strategic business goals, business benefits desired, program objectives, and business success factors is critical information needed by the team to ensure that the integrated program plan will be in alignment with company strategy.

Also carried forward from the earlier work cycles are the high-level requirements and a depiction of the whole solution. The high-level requirements are the foundation for the detailed requirements, and the whole solution becomes the basis for developing the program architecture.

Defining the Program Architecture

As detailed in Chapter 4, the program architecture begins with first visualizing the primary components that comprise the whole solution—the projects that directly contribute to the capability being created and delivered. The second piece of the program architecture consists of the support functions that enable the projects. PMI refers to these as "elements of related work outside of the scope of the discrete projects in the program" in its definition of a program.[7]

The program architecture shows which of the core components and enabling components of a program need to be implemented as projects, and which can be implemented as individual work efforts.

With the program architecture defined, the integrated program core team and extended team can be fully formed. It should be noted that selective membership of the IPT core team has already occurred to accomplish the work outcomes of the two previous program cycles.

Establishing the IPT Core Team

In order to create a comprehensive program plan, the program manager must first form an integrated core team. As described in Chapter 5, program team structure is a case of form following function. In other words, the makeup of the IPT core team membership is driven by the functions and disciplines that need to be involved in the program to provide the whole solution, as defined by the program architecture.

However, bringing a group of people together and assembling them as a team does not make them think and behave as a team. Members of a program team must view their work in terms of *we* instead of *me*. Meaning *we* must work together toward a common purpose that is defined by a set of common and agreed upon business goals. The program manager's job is to establish the common purpose and to inspire the team to work collaboratively to achieve the goals. It is the ability of the program manager to create a common vision and the team's willingness to adopt that vision that defines a group of people as a team.

A clearly defined common purpose is instrumental in removing ambiguity surrounding a program, and should answer four key questions: 1) what is the purpose of this team—the mission; 2) what does the end-state look like—the program strategy; 3) what do we need to accomplish to get to the end state—the program objectives; and 4) how will our success be measured—the business success factors. The job of the program manager is to create answers to these four foundational questions with input from the team members and stakeholders.

Defining the Program Scope

Program scope is determined by three key elements of program management: the program architecture, the program-level work breakdown structure (PWBS), and the benefits map. As stated earlier, the program architecture identifies the constituent projects of a program and the necessary functional enablers. The PWBS is a hierarchical decomposition of the deliverables and outcomes needed to realize the program objectives (Chapter 9).[8] If a project or functional support outcome is not included in the PWBS, it is not needed to achieve the business results desired and therefore should not be delivered as part of the program outcomes.

A common error that occurs when developing a PWBS is mixing both deliverables and tasks. A PWBS should only include deliverables or work outcomes, not tasks.

The PWBS does not eliminate the need for a detailed work breakdown structure for each of the constituent projects on a program. In practice both are required. The project managers use the PWBS as a guide for the scope of work for their respective project, and further break down each deliverable into the tasks and activities necessary to create it.

Projects within a program exist to help ensure a program is successfully executed, and as such, the final activity in defining program scope involves ensuring the project deliverables align to and support the program objectives. As part of the previous program definition cycle of work, program objectives were mapped to expected business results. During the program planning process, this activity continues by mapping project deliverables to the program objectives. The resulting *benefits map* establishes alignment of project deliverables, to program objectives, to expected business results (Chapter 9).

Creating the Integrated Program Plan

Creation of an integrated program plan requires a high degree of collaboration between the members of the IPT core team. Gone is the time when program leadership is authority-driven with work being performed through a series of directives. Today's era of program team leadership is a more diffused, collective team-based leadership, where directives and decisions are participatory, collective, and democratic. Success of the program manager is dependent upon how well he or she facilitates the alignment of interests, motivation, work activities, and collective outcome of the organizational parts.

With a good understanding of who is involved on a program and what will be delivered as established during the program structuring process, the program team is positioned to create the integrated program plan.

Four primary steps are involved in the development of an integrated program plan:

1. Documenting the detailed requirements
2. Developing project plans
3. Mapping cross-project interdependencies
4. Creating the integrated plan

Documenting Detailed Requirements: Creation of a good program plan is contingent upon how well the detailed requirements are defined and documented. If gaps exist in the detailed requirements, gaps will exist in the program plan. These gaps *will* get addressed, but normally later in the program in the form of scope changes. Likewise, low quality or nonspecific requirements *will* have to be redefined when someone tries to interpret them and perform real work against them. The later the reconciliation occurs, the more expensive it becomes.

The detailed requirements begin with the high-level business, technical, and customer requirements delivered at the end of program definition. Each member of the integrated core team is responsible for ensuring the requirements needed to perform their specialized functional work are included and adequately stated in the detailed requirements documentation. The detailed requirements must be clear, concise, and thorough, as they are the basis on which the project teams will create their work breakdown structures and project plans.

Once completed, the detailed requirements constitute the program scope and should be managed as living documentation. It is recommended that the program manager initiate a comprehensive stakeholder and peer review of the requirements to check that they are complete with respect to the information known about the program at the given time and that the quality level is in alignment with the quality goals and expectations set forth within the firm.

It needs to be understood that more will be learned about the program as time progresses, and that changes to the requirements may be needed during later stages of a program.

Initial Project Planning: Program managers serve in the capacity of project sponsor for each of the constituent projects on a program. As such, they will initiate the project charters and oversee the project planning process.

For the project managers, one of the realities of program management is that planning a project that is part of a program is strikingly different than planning an *independent* project. The work of the project manager working within a program construct is complicated by the fact that the project is a component of a larger effort that has many interdependencies. Because a project has to be planned within this larger context, multiple planning iterations may be required.

The first iteration is similar in fashion to planning that occurs on an independent project and begins with understanding the scope of the project. Once the detailed requirements are gathered and analyzed, each project team creates a project-specific work breakdown structure (WBS).

With their project WBS complete, each project team can begin developing their respective initial project plan. We use the term *initial project plan* intentionally, as a final project plan cannot be developed until all the cross-project interdependencies are identified and comprehended at the program level. Initial project planning ends with an understanding of the deliverables required for each project, the tasks and activities necessary to create each deliverable, and an estimate of time and resources to complete each deliverable. With this information at hand, the project manager is ready to participate in the next step—comprehending the cross-project interdependencies.

Mapping Cross-Project Interdependencies: At the project level, the project teams are concerned with development of their respective deliverables. At the program level, the IPT core team is concerned with the *interfaces* between the deliverables—which deliverables are needed by whom and when. The creation of the program map is the activity that defines the critical cross-project interdependencies (see Chapter 9 for details on developing a program map).

Each deliverable identified in the program work breakdown structure discussed earlier is displayed on the program map in the correct sequence and point in time. The use of arrows between deliverables depicts the interdependencies between project teams and their respective deliverables. This mapping of deliverables from one project to another helps the program team determine and fully understand the dependencies that exist on a program.

Besides the tangible benefit of understanding the cross-project interdependencies, the program mapping process has the intangible benefit of forging a change in mindset and frame of reference on the part of the project managers and functional specialists on the program. The change of reference involves moving from a project or function-perspective, to a program perspective. Instead of narrowly thinking of success in terms of delivering project outcomes, a broader view of program success in terms of collective success emerges.

Final Project Planning: The program mapping process provides each project manager with critical information that he or she needs to complete their final project plan. Most notably, an understanding of all the deliverables on the program, knowledge about who is dependent upon their team's deliverables and conversely who they are dependent upon, an understanding of the sequencing of deliverables over time, information about any program level constraints such as time, budget, or resources, and any program-level risks that may impact the work and outcomes of their project.

With this program-level information at hand, each project manager can again take a traditional approach to completing their detailed project plan. The program manager should set and communicate a common format and content expectation for each of the project plans. Primary elements that should be included in the project plans include the following:

- Project description
- Project objectives and success criteria
- Project deliverables
- Project schedule
- Project budget
- Team structure and resource profile
- Project risk analysis
- Tracking and reporting methods
- Change management methods
- Team communication strategy
- Project termination plan

To keep the work of the various project teams synchronized, the program manager should set clear milestone dates for project plan completion, and set up periodic reviews to ensure planning work is progressing, and to assist them in eradicating any barriers to successful completion. As changes to the program-level requirements and scope occur, the program manager should quickly communicate the changes to the entire IPT core team, so the changes can be incorporated in the project plans.

Integrating the Project Plans: Once the project plans are completed, the program manager can lead the program team through the creation of an integrated program plan. The program plan incorporates the work of all project teams and all other functional or discipline specific representatives that are part of the program team.

The program plan will vary by program type and situation but should include the following details:

Program Timeline. Using the output of the program mapping process as a starting point, the program team will determine the timeline of the program in the form of the master schedule. For most programs, it does not make sense to create a detailed activity-based schedule based upon the project schedules. The result is usually unmanageable or inefficient to manage. Rather, the program-level schedule should consist of an aggregation of the project and functional-specific deliverables in the form of milestones. The program manager can refer to the more detailed individual project schedules and plans as the need arises.

The use of timeline contingency is highly recommended. Inclusion of timeline contingency is a best practice that increases the probability of program success from a timing perspective. Contingency is a planned amount of time that is added to the estimated schedule to proactively account for errors, misjudgments, and unknown events (risks) that *will* occur. Contingency is derived from the risk management process and is based upon the various risk responses chosen (Chapter 7).

Resource Plan. Once the initial program timeline is developed, the project managers and the program manager can identify the resources needed to complete each task. The program resource plan is an aggregation of the project resource plans.

It is important to remember that resource estimates include both human resources and nonhuman resources. Nonhuman resources include supplies, materials, and capital equipment that people require to complete their work.

Detailed Program Budget. The program budget represents the organization's financial investment in the program. Much like the development of the program timeline, each project team is responsible for the development of their respective project budgets. The program manager works to aggregate the project budgets into an overall program budget.

The inclusion of budget contingency based upon the program risk assessment is a best practice that increases the probability of program success from a monetary or investment perspective.

The Program Strike Zone. The critical business success factors that were established in the definition cycle of the program can now be completed based upon the program planning work. Additionally, the key performance indicators for the program are established during the program planning cycle. A tool called the Program Strike Zone (Chapter 9) is effective for documenting both the business and performance success criteria against which the program will be measured.

Risk Assessment. A great deal is left unknown at the time a program plan is established. A thorough assessment of the potential problems that may be encountered that could impact the realization of the program objectives should be conducted and incorporated into the integrated program plan.

Each project manager should come into the program planning process with a set of project-specific risks that his or her team identified and assessed. The program manager is then responsible for collecting all risks and leading the integrated core team through an exercise to determine which of the risks are program-level risks and which are project-only risks (Chapter 7). The program risks are then categorized and prioritized by potential impact to the program.

As covered earlier, the risk response information should be utilized to estimate schedule and budget contingencies that are needed to protect the program success criteria in the event any of the risks come to fruition.

Validating the Program Business Case

As the business manager on a program, the program manager needs to ensure that the planning activities remain grounded in achieving the business benefits driving the program. The program business case is the program manager's guiding tool for ensuring that business alignment remains intact. Once the integrated program plan is completed, the program manager and team members need to validate that the program business case is still viable. This seems like an obvious step, but in practice it is a step that is often overlooked. It is crucial that the business case is periodically reevaluated to keep the program feasible from a business perspective. This is especially true at major decision checkpoints.

The Execution Readiness Decision

Having completed the outcomes required during the program planning process, the program team is prepared for the execution readiness business decision checkpoint. The program manager is responsible for presenting the integrated program plan and the validated business case to the sponsoring senior executive or the program governance board if one is established.

Normally, three decision options are available to the decision maker(s):

1. Do not approve the integrated program plan with direction to the program team to further refine the plan or the program business case.
2. Do not approve the program plan and terminate all further work effort on the program.
3. Approve the integrated program plan with direction to progress to the next cycle of work. Approval may come with sponsor and stakeholder redirection.

If approval is achieved, the program team is given authorization and commitment of resources to execute the integrated program plan. During this next work cycle, the program team prepares for the capability release decision by engaging in program execution and creating the capability in accordance with the program business case.

DECISION CHECKPOINT: CAPABILITY RELEASE

With an integrated, cross-organizational program plan in place and resources and funding committed to execute the program, the next major business decision ahead of the program team is the capability release decision.

The intent of the capability release decision is to evaluate the readiness of the capability to be released to the market or introduced into an organization. Also under evaluation is the readiness of the program team to support the capability and to sufficiently manage the change that the capability will introduce.

The cycle of work ahead of the capability release decision, commonly referred to as program execution, is focused on accomplishing the design and development work necessary to create the capability, integrating the work output to ensure the whole solution is developed, preparing for the release of the capability, and ensuring the program is managed to the intent of the business case.

Managing the Program Execution Cycle

The program execution cycle is arguably the most anticipated phase of a program. It is when an intangible concept becomes a usable, tangible asset for realization of the business goals. It is also the part of the program in which the quality of the integrated program plan and the ability of the program manager to lead will be put to the test.

The job of managing a program during the execution process can be challenging due to a couple of natural factors. First, the size of the program team and the number of cross-project interdependencies to track and manage grows rapidly between planning and execution. Figure 6.2 illustrates this phenomenon.

The program profile shows that many elements of the program, such as staff size, budgeted dollars spent, and number of interdependencies, are at their highest levels and peak during program execution. These factors need to be closely managed both within the projects by the project managers and across the projects by the program manager to ensure that the program stays in alignment with the business objectives. The challenge for the program manager is to remain at what we call the 10,000-foot level of a program and focus on managing the cross-project collaboration, while empowering the project managers at ground level to focus on the detail necessary to accomplish their deliverables. Program managers

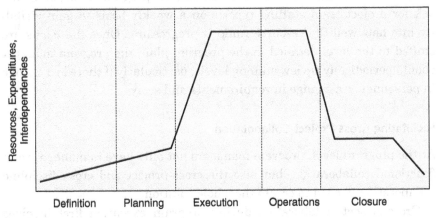

Figure 6.2 Common program profile.

cannot allow themselves to be pulled to ground level for very long, or cross-project collaboration will begin to suffer.

The second factor that complicates the management of program execution is that all program stakeholders suddenly become aware of a program's existence, seemingly overnight. It is as though many stakeholders do not pay attention to a program until it becomes *real* in their eyes, meaning that until a plan has been approved and resources are working to implement it, a program does not really exist. When this happens, the program manager has to spend more time away from the program team to manage the expectations and inquiries of the stakeholders. The need for a stakeholder strategy is again prevalent (Chapter 7).

Staffing the Extended Program Team

Upon approval of the integrated program plan, one of the first activities to take place during program execution is the selection and assignment of all human and nonhuman resources to the program. The program manager is responsible for ensuring that the IPT core team is fully staffed, while the project managers are responsible for staffing their project teams.

The program manager is not responsible for selecting the resources for the entire program. Rather, he or she ensures each project manager has the ability to staff their project team with the number of resources required, as well as fulfill critical skill requirements. If gaps exist in these areas, the program manager should work with functional managers, executive managers, and human resources to fill resource gaps. If critical gaps cannot be filled, changes to the integrated program plan may be required. During the early part of program execution, the program manager should ask for project team staffing reports on a weekly basis to gain visibility into how well the resource ramp is progressing. Once the teams are staffed to the level detailed in the program plan, the program manager should periodically review staffing levels, particularly if there is a change in personnel or a change in requirements and scope.

Facilitating Cross-Project Collaboration

At the program level, program managers not only have to manage cross-functional collaboration, but also the cross-project and cross-discipline collaboration needed to create the whole solution.

Cross-project collaboration during program execution first involves facilitating the highly complex network of project interdependencies in the form of deliverables. The program manager is responsible for ensuring these cross-project interdependencies remain synchronized and coordinated. The program map, which was created and matured during program planning, is one of the most helpful tools for managing the implementation of the deliverables (Chapter 9). The program manager can use the program map as a rolling wave execution tool, focusing the program team on cross-project deliverables due within a four week rolling window, for example. This practice will facilitate the cross-project discussions on status, risks, and issues that need to take place between the project teams.

Managing Program-Level Risk

Team-based risk management is a powerful cross-project collaboration practice. Through team-based risk management, the program team has a greater knowledge of the risks to the program, which helps them think of risk in a broader perspective—in terms of the impact of risk outside of their specialty domain and how cross-project risks must be managed collaboratively.

The program manager is the risk management advocate on the program and is responsible for evaluating, managing, and communicating the overall risk of the program. The IPT core team is responsible for identifying and tracking the key program-level risks across the elements of the program, while the project managers are responsible for identification, assessment, and management of all program and project risk events specific to their discipline. The project managers are also responsible for bringing new risk events identified within their project to the program level for cross-project evaluation and inclusion in the program risk portfolio.

By utilizing the program risk management process similar to the one described in Chapter 7, along with tools such as a P-I matrix (Chapter 9), the program manager can effectively manage the overall risk of the program throughout the execution process.[9] Program risk must trend downward over time, as cumulative risk is an indicator of program health. As the program approaches the end of execution, program risk will be a deciding factor on whether to release the program's capability.

Managing Change

Change management is a critical practice necessary to control the strategic direction and scope of a program. Changes that impact programs can originate at the business, program, and project level. The organization's governance policies and procedures will normally specify which responsible managers must be consulted and involved in change-based decisions depending upon the impact on the program's benefits objectives. Generally, project-level change focuses on realignment to the program plan. All other types of program changes need to be assessed as to their impact on the business benefits and program objectives.

Historically, uncontrolled change has been a primary cause of program teams' failure to meet their intended goals.[10] During program execution, rapid change is common. Shortly after a baseline plan is approved, changes begin to occur. It is important that the program manager establishes a robust change management process to evaluate change benefit versus cost to the program.

Implementing Program Governance

Program execution never occurs exactly as planned, creating a need for effective program tracking and control practices on the part of the program manager. A program manager can neither eliminate changes in the environment nor avoid all errors made during program definition and planning, but he or she *can* use good tracking and control techniques to minimize the impact of these factors on the program objectives. Tracking progress of work consumes a large portion of a program manager's time and mind share during program execution.

Whenever possible, the control of the program should be administered within the team. This is a function of the empowerment from senior management to the program manager. Trust and credibility will be further enhanced in the eyes of management when they observe program managers and their teams properly monitoring progress and performance, and taking the appropriate corrective action when the need arises. Program tracking and control consists of the following elements:

- Determining what elements of the program to track
- Deciding what metrics and tools to utilize
- Managing program deviations

What to track: Deciding on what elements of the program to monitor is the foundation of good program governance. Many of the important and critical elements of the program to track will have been decided between the sponsoring executive and the program manager during identification of the critical success factors for the program. This is an area in which a program manager can get buried in the details of the program, if not careful. The program manager should focus on monitoring program-level elements such as major program milestones, overall program budget, cross-project risks, and changes to the program success criteria. He or she should then delegate the detailed tracking and control to the project managers. Project managers will track the progress of the critical elements associated with their functional specialty, with guidance from the program manager on standard elements that he or she wants tracked on all projects.

Metrics and tools: Selection of the program elements to be monitored will influence the metrics used to measure progress, and the tools used to collect the measurements. The same process applies to the elements that will be monitored on each of the projects. The program manager, however, needs to drive standardization of metrics and tools across the project teams.

The business success factors and the KPIs are the fundamental metrics of the program and are fully defined at the completion of the planning phase. During program execution, the program manager should monitor the progress of the program toward achievement of the success factors, as they represent the targets for successful program completion. The program strike zone is an excellent tool to identify the business success factors of a program and to help the organization track progress toward achievement of the key business results desired (Chapter 9).

The program dashboard is a tool that highlights and briefly describes the status of a program based upon the primary metrics selected. It can be used to address a specific program or by senior management to summarize multiple programs within an organization. On a specific program, it provides status information relative to progress toward achievement of major business goals. For executive managers, dashboards can be used to summarize multiple programs underway in order to provide a quick understanding of the status of all programs.

Managing variances: The purpose of program metrics and tools is to monitor the progress of the program team in execution of the program as laid out in the integrated program plan (Chapter 8). If effective, the metrics and tools will give the program manager an early indication that deviations in progress as compared to the program plan have occurred.

However, it is not enough to simply detect a variance between actual performance and planned performance. The important aspects of managing the deviation involves understanding what it means, what caused the deviation, and then determining what to do to correct it.[11] The program manager may need to adjust project team activities or resources to bring performance back into alignment with the plan. He or she may also need to adjust the integrated program plan to compensate for errors or unexpected changes.

Preparing the Capability Release Checklist

As program execution nears completion, the program team must begin preparation for release of the capability that the program has produced. The capability release plan is part of the overall integrated program plan and is completed during the planning phase. In preparation for the capability release decision checkpoint meeting at the end of program execution, the program manager will lead the team through preparation activities, including the completion of the capability release checklist.

The checklist is a summary-level listing of the primary release activities, with a complete or incomplete indication shown for each activity. Figure 6.3 illustrates this type of checklist, which identifies a set of activities and completion status. Keep in mind that each program is unique,

Release Item	Completed
Customer readiness	☒
Customer support readiness	☒
Production readiness	☒
Distribution channel readiness	☐
Supply chain readiness	☒
Sales force training	☐
Demonstrations planned	☒
Public announcements planned	☐

Figure 6.3   Example of capability release preparation checklist.

and therefore each will have its own set of activities. The checklist will be a primary item for evaluation during the capability release decision meeting.

Many organizations have found these release checklists so important that they have implemented them for preparation at each decision point for the program team.

The Capability Release Decision

The final step in program execution is to conduct the capability release decision meeting. The program manager typically presents the status to the executive decision-making body in the form of a release proposal. As in every business decision checkpoint meeting, the business case should be updated and reviewed to ensure that the program is still viable from a business perspective prior to moving to the next cycle of work.

Normally, three decision options are available to the decision maker(s):

1. Do not approve the capability release proposal with direction to the program team to repeat the appropriate elements of program execution.
2. Do not approve the capability release proposal and terminate all further work effort on the program.
3. Approve the capability release proposal with direction to progress to the next cycle of work. Approval may come with sponsor and stakeholder redirection.

Successful completion of program execution should culminate in the full commitment of funds and resources to release the capability into the market or organizational environment and to begin program operations. The program will remain in operation until the appropriate time to evaluate the need for program closure.

DECISION CHECKPOINT: PROGRAM CLOSURE

In a normal scenario, program closure occurs after the program capability has been in the market, user environment, or in an organization for an appropriate period of time to realize the business benefits intended. However, it should be noted that the program closure decision can also be made at any point in the program cycle if the business environment or program performance has changed to the point where the program is no longer needed or the benefits are no longer attainable.

In either scenario, the final business decision—the program closure decision—can be made to formally stop all program activities and reallocate resources and funding to other programs in the portfolio.

The cycle of work ahead of the program discontinuance decision is centered on a number of primary objectives: releasing the program capability, assessing benefits realization, capturing program knowledge, and preparing for the program closure decision.

Managing the Program Operations Cycle

When a program transitions into an operational mode, the release of the program output has been accomplished and detailed tracking of business benefit realization begins. It should be understood that full benefits realization may take significant time. Since programs require an investment in money and resources, it makes sense, therefore, to close a program as soon as it is reasonable. In some cases, long-term benefits assessment should be based upon the *probability* of achieving a benefit.

If the program output is delivering business results as anticipated, or better than anticipated, efforts may be taken during program operations to improve the capability to adapt to emerging changes or needs in order to extend its operational life.

Before benefits realization or capability improvement can happen, however, the release of the program output into the market, into the customer environment, or within an organization has to occur.

Releasing the Program Capability

This step involves formally releasing the output of the program into the operational environment of the business. For a product development program this means beginning to produce the product in saleable quantities. Production processes, supply-line processes, and distribution channels are all turned from development to regular production status. Additionally, the order and fulfillment process begins to take formal orders and fulfills those orders to customers.

For a service or infrastructure program this means formally moving the capabilities from the development and test environment to the operational environment and integrating the capabilities with all other systems within the organization. At this stage, end users are able to exercise the capabilities to their full extent.

For a change transformation program this means introducing the change into the market or organizational environment. Operational support, training and mentoring, and discontinuance of old processes and capabilities have to be carefully managed.

As customers or clients begin to use the capability, the support team begins to respond to requests for assistance and information. All procedures developed by the team now become operational, and customer support metrics are collected.

Likewise, operational support for the capability begins. This may include such things as performing maintenance procedures, fault isolation, testing and repairs of failed units, or technical operational assistance at the customer sites. Many organizations add the new capability to their ongoing sustaining group to ensure that needed quality and process improvements are managed as needed.

The tracking and analysis of the early capability support metrics are critical in effective management of the capability release. With any new capability, defects or other problems may show up once the customer begins using it. Besides ensuring the customer's problem is resolved as quickly as possible, it is also important to understand the cause of the defect or problem and fix it so it does not reoccur. This field data is also of great value to the architects that may be defining the next generation of the capability.

Assessing Benefits Realization

The responsibility for managing the program business continues through program operations as it is during this stage that the business goals from which the program was spawned should be realized. For example, the ROI predicted in the business case should be achieved, the cost savings targets should be met, market segment share gains should be realized, or a new market or market segment should be opened to the company.

Benefits realization during program operations involves ensuring that the business case supporting the program has been achieved and that the program objectives have been realized. Achievement of the business case ensures the realization of the anticipated business results, while realization of the program objectives gives an indication that the strategic business goals of the enterprise have been at least partially attained.

Program governance during operations should focus on ensuring that both the business case and the program strike zone, which contains the program objectives, are continually monitored for progress against business benefits.

Program Knowledge Capture

A program *retrospective* should be conducted to assess, document, and discuss the successes and recommended improvements for future programs. Collecting and acting upon program key learnings is a necessary step in becoming a learning organization and, in doing so, strengthening the foundation for the next generation of programs.[12] When collecting data, include inputs from the IPT core team, extended team, sponsors, and other stakeholders affected by the program.

Findings should be documented and communicated to key stakeholders in the organization, including the program team, other program managers, functional managers, and executive management. The information should also be used to build a key learnings report and identify process improvement initiatives, as appropriate. Senior managers should understand and convey that the retrospective exercise is directed toward continual improvement of the organization, rather than letting it come across as finger pointing or assigning blame if all did not go well on a program. A manager does not want to limit or minimize the open flow of needed information.

We present the program retrospective at the end of the program, but best-practice companies typically perform a program retrospective following each major business decision checkpoint that focuses on the prior work cycle. The benefit of this approach is that the information is fresh, more comprehensive, more focused on a single work cycle, and learnings are fed back into the organization more quickly.

The Program Closure Decision

The final step in the program operations process is to conduct the program closure decision meeting. The program manager typically presents a closure proposal to the executive decision-making body of the organization. The decision to close a program is heavily based upon the program's performance against the business case and assessment of whether or not the critical business factors have been met. This is generally the point in a program where a judgment will be made on the level of success of a program.

Normally, two decision options are available to the decision-making body:

1. Do not approve the closure proposal with direction to the program team to continue program activities.
2. Approve the closure proposal with direction to perform program closure activities.

Successful completion of program operations should culminate in the achievement of the business benefits from the program and the start of the final cycle of work—program closure.

Program closure begins with program planning. The closure plan should be developed as part of the integrated program plan and executed following program operations. When this occurs, the program manager leads the team through all activities necessary for closing the program.

Exact elements of the closure plan vary from program to program, but in general program closure involves stopping all projects, reassigning resources, managing the capability phase-out or hand-off, and capturing and sharing program knowledge.

The program manager needs to be aware that at times there is reluctance on the part of some program team members to discontinue their work. This is an emotional reaction to upcoming change that the program manager should be aware of and willing to provide the necessary leadership to support members of the team.

Program Management in Practice: Program Management Goes to University

Contributed by Kathy Milhauser

It was 2009 and Linda Kramer had to make an important decision. Linda was the director of a number of degree programs at a major university headquartered in the Pacific Northwest. The university leadership had already decided to enter the competitive market of graduate studies with a new degree, a Master of Science degree in project management. As the director of the project management degree program, Linda was responsible for planning and designing the curriculum for this program and administering its multi-modality operational delivery. The decision she had to make was based largely on one complex question:

How will I design the curriculum in a way that competes effectively in an intense and rapidly growing market while achieving desired business goals relative to ROI, growth and market positioning, customer satisfaction, and accreditation-approved quality?

This question is a common one in higher education institutions around the globe today. Linda knew the question was a common one. She also knew that her answer needed to be unique to "fit" her organization's specific situation, strategy, and goals.

Lacking Curriculum Coherence

The university had experienced rapid growth and success in financial position and market position over the prior decade. The success was largely a result of international partnerships and expansion into online course delivery. The success was certainly not due to the university's new product development methodology.

While the university leadership was adept at discerning market opportunities, it did not have a standard process for taking a strategic initiative from idea to operational reality. Linda was informed upon hire that the university had no disciplined or standard framework for new curriculum development, other than a focus on meeting and maintaining accreditation standards. In speaking with other directors of degree programs, Linda quickly realized the painful truth of these words; most of their development had been ad hoc and success or failure was largely due to the efforts and business savvy of the directors themselves. That said, the university was clear on its decision—the new Master of Science degree in project management was already being marketed for a fall semester launch, less than a year away. The primary goal was to have a cohort (i.e., a group of people registered for the curriculum) of at least 20 students for this launch. Thereafter, a goal of starting three cohorts per year was planned and already part of the university revenue projections. Linda needed to meet these targets.

A third goal of the university was to work on curriculum integration. A theme emerging from student feedback was a lack of coordination and progression of coursework. The university wanted the faculty to work together more closely regarding coursework activities (readings, cases, assignments) to create greater integration and synergy, which then would contribute to enhanced learning by the students. These improvements were expected to create higher rates of student retention.

Developing a Whole Solution

The organizational situation and ad hoc product development practice are not unique to this university. The fact that directors are responsible for delivering graduate degree programs (like the Master of Science in project management for which Linda was responsible) at this university is similar to other academic organizations. University leadership forecasts revenue projections and relies on recruiters to fill cohort targets and relies on directors to prepare and execute curriculum based on their best ability.

Linda's best ability was recognizing the need to manage the Master of Science in project management degree program as exactly that, *a program*. Prior to working at the university, Linda's experience was in planning and delivering training and workforce development initiatives for a global leader in the sports footwear and apparel market. Linda was steeped in new product development practices and in using program management as the discipline to manage her work.

With the ad hoc culture of the university, using program management was going to be unique, but she saw it as a great opportunity to overcome a major concern of university administrators and their customers (students and the companies where the students work)—to better integrate coursework so that it is a holistic, seamless product rather than a series of loosely aggregated silos of classes. After many conversations with university leaders, Linda was convinced that following a program management approach provided the highest probability that the degree program would be market-driven, rather than faculty-driven, and ready in time for launch.

Program management is about the business. As such, program management is market-driven and focused on customer demand. Product development organizations often employ a life cycle model that begins with understanding markets, competitors, and customers. The industry research and business analysis efforts yield detailed requirements for inclusion in the product's specifications.

When applying the product development model in higher education, the students and, perhaps more importantly, the companies that hire and employ the students upon graduation fill the role of the "customer." Their needs are converted into "specifications" and "requirements," which are then used in planning, designing, and developing the product, which in this case is the whole curriculum. So, in the case of higher education, a product focus suggests the needs of the hiring company drive the curriculum design and delivery mode, which in turn drives the process for developing the student into a competent, highly skilled professional equipped with relevant abilities for success in the workplace.

Linda knew that aligning the curriculum with market demand in a rapidly evolving workplace was a relevant and important role for higher education degree programs, and that the education industry was more similar than different than product development in other industries. While this approach is not new, it is still unique in practice. For example, many institutions "teach" to the technical training competencies of their faculty rather than focusing curriculum on customer need. This is commonly referred to as Intuitive Curriculum Design (ICD) and varies greatly from market-driven design, which is commonly referred to as Instructional System Design (ISD).

ICD is an "inside-out" (university-to-market) approach that relies on individual faculty to design, develop, and execute (teach) individual courses based on their expertise. As such, the ICD approach is course specific, not whole solution specific. The ISD, on the other hand, is an "outside-in" (market-to-university), product-based approach and used to plan, design, develop, and evaluate training on the whole solution. It was based on these benefits that Linda promoted the use of program management using ISD protocols at her university.

Employing a Life Cycle

Linda's process for development and management of degree programs is illustrated in Figure 6.4. The map includes four major phases of work: market, institution, program, and course. (Note: The use of "program" in the third phase is not here to confuse the reader, but rather to note the Master of Science in project management program specifically.) At the university they use the word "program" in two ways: first, to describe any degree granting program or curriculum that culminates in a degree, and second, to describe the complex undertaking of work that is associated with multiple, interdependent projects and linked to an organization strategy and set of business goals.

As the director of the program, Linda was responsible for the whole solution. She involved a small core team of faculty to develop the curriculum to meet market needs as defined by industry stakeholders (practitioners, hiring managers, accrediting professionals, and more).

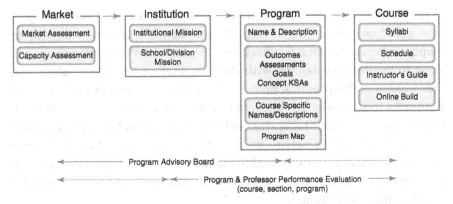

Figure 6.4   Curriculum product process map.

In the "Market" phase of the process, the market need, capacity, and characterization is analyzed. This involves continuous engagement with members of business industry currently facing challenges similar to those that graduates from the university will soon be facing. Leaders in industry who are responsible for hiring are also very valuable partners during this market analysis phase, as well as throughout the development process, especially as courses and degree programs are evaluated. Feedback can be reciprocating in that the business practitioners and hiring managers can provide detail regarding market changes, demands, and expectations, and the degree program and course effectiveness can be fed back to the partners.

The "Institution" phase of the process was where alignment to the university's mission and vision, as well as the mission and vision of the specific school or division, was maintained. This also supports the alignment to accreditation standards, which inform and support alignment of strategic planning processes in the higher education landscape. Like hiring managers in the market phase, accrediting stakeholders are very valuable partners during this phase as well as throughout the development process.

The "Program" phase of the process was where the Master of Science in project management was defined; outcomes articulated; key knowledge, skills, and abilities determined; and a holistic curriculum was developed to guide the design and development of coursework. Subsequently, the "Course" phase of the process was where specific syllabi, schedules, instructor lesson plans and guides, and online and classroom course content was built. It was at this phase of work that cross-course integration points were intentionally established.

Linda knew, and expressed to her administrators, that how this process would be managed is as important as how it was designed, with an emphasis on continuous feedback and measurement of performance relative to business goals, and with the customer engaged throughout the four phases of the program process. Thus, the notations at the bottom of Figure 6.4 indicate the involvement by Program Advisory Boards, Accreditation and Assessment processes, and Program and Professor performance evaluations. Business metrics, program metrics, and operational efficiency metrics were identified and used to maintain effectiveness of the program to launch the master's curriculum as well as to measure the operational effectiveness after launch.

The intent of the model was to assure university administrators that alignment and integration would occur between market needs, business strategy, curriculum development, the course training effectiveness, and what the student ultimately learns.

In Retrospect

Linda, working with the university administration, recommended an integrated, outside-in, outcomes-based approach to address the high demand knowledge and skill areas identified in project- and program-oriented businesses. For quality assurance, she recommended alignment to accrediting standards as well as established learning theories to ensure that the knowledge, skills, and abilities that were embedded in the curriculum were the right ones (as defined by the customers) and were effectively transferred to the students to enhance their value in the market.

Finally, Linda advocated a product development approach that could help higher education institutions satisfy other stakeholders (accrediting bodies) by aligning curriculum to industry needs as well as professional standards. She explained this importance to university administrators for two reasons: 1) to keep them in the loop regarding her approach to the new program, and 2) to informally educate them on the value of a program management discipline in their organization.

Currently, the university's Masters of Science degree in project management has been successfully implemented and meeting the needs as measured by hiring organizations. The program objectives were achieved relative to meeting the launch date and the number of cohorts. The program management discipline and approach that Linda helped develop for the university continues to expand today.

ENDNOTES

1. Martinelli, Russ, and Steve Graykowski. "Agile Program Management: Separating Practice from Myth." *PM World Today,* June 2010, Vol. 12.
2. Doran, George T. "There's a S.M.A.R.T. Way to Write Management Goals and Objectives." *Management Review,* November 1981, pp. 35–36.
3. Project Management Institute. *The Standard for Program Management,* 3d ed. Newton Square, Penn., 2013.
4. Office of Government Commerce. *Managing Successful Programmes,* 3d ed. Norwich, UK, 2007.
5. Young, Ralph R. *Project Requirements: A Guide to Best Practices.* Tysons Corner, Vir.: Management Concepts Press, 2006.
6. Project Management Institute. *The Standard.*
7. Ibid.
8. Sasghera, Paul. *Fundamentals of Effective Program Management: A Process Approach Based on the Global Standard.* Plantation, Fla.: J. Ross Publishing, 2008.
9. Smith, Preston G., and Guy M. Merritt. *Proactive Risk Management.* New York: Productivity Press, 2002.
10. Archibald, Russell D. *Managing High Technology Programs and Projects.* Hoboken, N.J.: John Wiley & Sons, 2003.
11. Milosevic, Dragan Z. *Project Management Toolbox.* Hoboken, N.J.: John Wiley & Sons, 2003.
12. Lewis, James P. *Fundamentals of Project Management.* New York: AMACOM Publishers, 1997.

Part III

Program Practices, Metrics, and Tools

Part II explained how best to navigate programs to deliver business benefits. The three chapters in this section detail the practices (Chapter 7), metrics (Chapter 8), and tools (Chapter 9) used to effectively and efficiently manage programs and program teams.

We do not advocate a prescriptive approach to implementing a set of program management practices. We know from experience that each organization is unique and needs to implement the practices that best fit its business processes, culture, and level of program management maturity. There are, however, a number of core program management practices that we see being used consistently across business sectors and levels of organizational program management maturity. In Chapter 7, "Program Management Practices," we present this set of core program management practices and explain how they are commonly used to manage a program to success.

In addition to practices, the use of program metrics can increase program management efficiency and benefits realization. Chapter 8, "Program Metrics," clearly defines metrics and details their use as a powerful resource within programs and organizations. We explain why program management metrics are needed, the various types of metrics, and how to determine which program metrics are important based on organizational strategy.

We wrap up this section with Chapter 9, "Program Management Tools." Program management tools support the practices and various processes used to effectively manage a program. They are enabling devices for the primary players on a program. The tools presented in this chapter are a key subset of tools that we have found to be most impactful and widely implemented within best practices organizations. Additional tools and tool templates can be found on the Program Management Academy website: http://wiley.programmanagement-academy.com.

Chapter 7

Program Management Practices

The previous chapter described how a program can be managed through a series of business decision checkpoints and work cycles. Rather than advocating a prescriptive approach, we presented a framework in which an organization can nest its preferred processes and methodologies.

In like manner, we do not advocate a prescriptive approach to implementing a set of program management practices. We know from experience that each organization is unique and needs to implement the practices that best fit its business processes, culture, and level of program management maturity.

There are, however, a number of core program management practices that we see being used consistently across business sectors. In this chapter we present this set of core program management practices and explain how they are commonly used to manage a program to success.

BENEFITS MANAGEMENT

Benefits management is about realizing the business results desired from the investment in a program. In a recent blog, a gentleman made the case that the primary directive of program management was to achieve the program objectives.[1] Another person joined the discussion and made the case that program management was instead about achieving business results and *benefits management* was really the primary directive of program management. We are in this person's camp. This extension in thinking was encouraging to observe and demonstrated to us that people understand the true value of program management.

Benefits management at the program level is about management of the business goals and achievement of the business results driving the need for the program.

Benefits Management Strategy

Creating a benefits management strategy in effect involves identifying the specific business results that a program must deliver to directly support the strategic goals of the business. As we explained in Chapter 6, this occurs early in a program and is central to the program definition work cycle.

The benefits management strategy is the foundation of the business case for a program. It defines *how* a program will contribute to the realization of a firm's strategic goals if the program is funded and properly executed.

In Chapter 3 we described how, for program-oriented organizations, program management is part of the business engine of the firm. The benefits management strategy is the keystone element that ties the strategic management, portfolio management, and program management components of the business engine together as a coherent management system.

Benefits Mapping

Creating strategy is critically important, but can be a wasted effort if the setting of strategy is not followed by successful execution of strategy. This is true as well for benefits management. The process of benefits mapping ensures the execution outputs from the constituent projects of a program support the benefits management strategy and the business results anticipated.

First created during program definition and then refined during program planning, the benefits map establishes direct alignment of project deliverables or outcomes to the program objectives, and then to the expected business benefits and strategic goals of the firm. Creating a benefits map is detailed in Chapter 9.

Benefits mapping and the resulting benefits map are necessary elements for defining the program scope and the benefits realization plan, both of which are core components of the program business case (Chapter 6).[2]

Benefits Tracking

The process of benefits tracking involves periodically examining the progress of a program toward achievement of the expected benefits. In practice, firms have found that it becomes increasingly more difficult to terminate a program the longer it is in existence.[3] If the critical business success factors are properly defined and consistently reviewed as part of the program governance system, the means to objectively evaluate the value of a program is available to the program sponsor and governance board at any time during the life of a program. More importantly, the sponsor and board have the ability to stop or realign a program to the benefits desired if changes to the environment or program make it necessary.

Program managers should ensure there is a specific distinction between the business success factors for a program and the key performance indicators. Both are important for tracking progress of a program. But a key performance indicator such as *performance to budget* may not define program success in the way that a business success factor such as *$200,000 in sales force expense reduction per year* does.

Benefits Realization

As pointed out in Chapter 6, there is often a delay between the release of the program capability and the full realization of the business benefits resulting from the use of the capability. Effective program management means diligent benefits management through focused attention on achieving the program business case. It is necessary to spend the time and effort to get the business case *right* and to gain support by the key stakeholders. Managing the program to the intent of the business case and delivery of the program output become the means to realize the business benefits desired.

Benefits Management Practices

Benefits management is the foundation of program management, and for this reason a program should be defined by the benefits it delivers. Benefits management runs throughout the life of a program, from initial ideation to program closure and beyond. Benefits management begins with business benefit identification during the program definition work cycle when specific business results needed to attain the strategic goals of the program are identified. As part of this process, the business success factors are documented and validated by key program stakeholders.

Key to benefits management is the development of the program business case that supports an investment decision. The potential business benefits to be gained are weighed against the investment cost of implementing a program.

Execution of a program is guided by the business success factors contained in the business case that keep a program aligned to the business benefits desired. Ultimately, program closure occurs when it becomes clear that the business benefits have been realized, are well on the way to being realized, or in the case of termination, cannot be realized. Program success is assessed against the attainment of business benefits and realization of the strategic business goals.

STAKEHOLDER MANAGEMENT

A stakeholder is commonly defined as anyone who has a vested interest in the outcome of a program. More importantly, for a program manager, a stakeholder is anyone who can influence, either positively or negatively, the outcome of a program. This includes people and groups of people both inside and outside the organization. Stakeholder management is a process with which a program manager can increase his or her acumen in managing the political, communication, and conflict resolution aspects of his or her program to ensure a positive outcome.[4]

In Chapter 1 we introduced the concept of the program management continuum that identifies various ways in which program management is implemented and practiced in organizations. As the program management function within an organization grows stronger and more influential (moving from left to right on the continuum), the more important effective stakeholder management practices become in order to successfully manage programs on a consistent basis.

This is due to the fact that as an organization transitions from a project-orientated to a program-oriented business, the accountability for success relies more and more on the program manager—a person who has limited positional power within the organization, yet still owns the responsibility for program success. In addition to empowerment from senior leadership, a program manager must work to earn cross-organizational empowerment from the building of strong relationships and successfully influencing key stakeholders.[5]

Stakeholders are many and varied on a program, and come to the table with a variety of expectations, opinions, perceptions, priorities, fears, and personal agendas that many times are in conflict with one another.[6] The challenge, therefore, is to find a way to *efficiently* manage this cast of characters in a way that does not become all consuming. Fundamental to efficiency is being able to identify and separate the highly influential stakeholders and then create and execute a stakeholder strategy that can strike a balance between their expectations and the realities of the program.[7]

Stakeholder Identification

Effective stakeholder management begins with identification of all stakeholders associated with a program. To begin with, the list of stakeholders should be comprehensive in order to cast a wide net over all players who may have a vested interest in the outcome of a program.

Tools such as a stakeholder map are common, and can be effective in helping a program manager identify the various stakeholders. The stakeholder map should include internal stakeholders such as executive managers, program governance board members, department or functional managers, support personnel (accounting, quality, human resources), and the program team. External stakeholders should also be listed and may include contractors, vendors, regulation bodies, service providers, and others. The objective of stakeholder identification is to include anyone who *might* have influence on the outcome of the program.

Stakeholder Analysis

Stakeholder analysis begins with the categorization of stakeholders into the logical groups that they belong to. Such categories may include senior sponsors, executive decision makers, team members, and resource providers to name a few. It is important to realize that some stakeholders may belong to multiple groups. The intent of stakeholder categorization is to bring structure to the stakeholder list based upon common interests in the program.

With categorization complete, stakeholder analysis turns to determining what type of influence each stakeholder has on the program, such as decision power, control of resources, or possession of critical knowledge, and their level of allegiance to the program. In other words, does the stakeholder prefer the program to succeed, to not succeed, or is he or she indifferent about the program?

Being able to determine the influential, or primary, stakeholders is critical to the program manager because he or she will not have time to engage with all program stakeholders. To effectively build relationships with the right stakeholders, the program manager must flush out who the primary stakeholders are, understand what level and type of influence they have, and determine how they feel about the program. A stakeholder analysis tool such as the one shown in Figure 7.1 can be very useful.

Now, let's be honest. Widely sharing the information shown in Figure 7.1 may be a career-limiting move. Detailed stakeholder analysis information should be considered confidential and tightly controlled by the program manager. Why do we say this? Because true and honest stakeholder analysis brings to the surface the realities of corporate politics and unique biases that have to be managed properly; usually that means actively, but delicately.

Many program managers have told us that they do the analysis in their head, which is good. However, when they hold the information in their heads, it tends not to be used consistently or correctly, which is to sort the primary stakeholders from the remainder of the stakeholders and to develop a concise stakeholder strategy. Performing the additional step

Stakeholder Name	Relationship to Program	Level of Influence	Allegiance to Project	Resources					
				People	Money	Material	Facilities	Knowledge	Decisions
Sue Williams	Sponsor	H	⇧		X			X	X
Ajit Verjami	Functional Manager	H	⇩	X		X			X
Steven Cross	Software Vendor	L	▭	X			X	X	
Lynda Donovan	Core Team Member	M	⇧						X

Influence Key:
L = Little or no influence
M = Some influence
H = Considerable influence

Allegiance Key:
⇧ Positive allegiance
▭ Neutral allegiance
⇩ Negative allegiance

Figure 7.1   Example of stakeholder analysis tool.

of documenting the stakeholder analysis information, either physically or electronically, helps to maintain history and clarity in developing the stakeholder strategy.

The Stakeholder Strategy

Stakeholder analysis is about sense-making. This means understanding the significance of the information gained about the various program stakeholders. Significance of the information is then used to develop a strategy for engaging and managing the *right* set of stakeholders. The right stakeholders are those who have power and influence to affect the outcome of the program.

Most of the literature on stakeholder management immediately classifies the primary stakeholders as those with both high power and strong allegiance to the program. This is significantly inaccurate because it leaves out the most potentially dangerous stakeholders: those with high power and negative allegiance to the program. These people also need to be considered primary stakeholders. Figure 7.2 helps to illustrate why. Stakeholders with high power and negative allegiance fall in the category of significant engagement required.

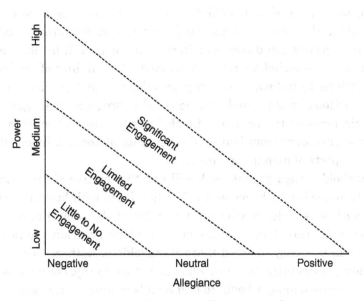

Figure 7.2   The power/allegiance grid.

If a program manager uses the power/allegiance grid to map their stakeholders, a core stakeholder strategy will begin to emerge. The strategy should consist of a communication and action plan for each of the primary stakeholders. It should also keep the program advocates engaged, describe how they can be used to influence others, and plan how to win over or neutralize the stakeholders who are not current advocates.

The stakeholder strategy should consider the following aspects:

- What is wanted or needed from each stakeholder?
- What is the message that needs to be delivered to each stakeholder?
- What is the best method and frequency of communication with each stakeholder?
- Does the strategy reflect the interests and concerns of each stakeholder?

The stakeholder strategy is not a static exercise or event to be performed once on the program. Rather, it needs to be both continuous and iterative throughout the life of a program to accommodate changes in program stakeholders and changes in attitudes and needs of the stakeholders.

Stakeholder Engagement

The primary goal of stakeholder engagement activities is to establish stakeholder alignment to the strategic goals, intended business benefits, program objectives, and success criteria of a program. It involves putting action to the stakeholder strategy through the building of professional relationships to influence for program advocacy and to monitor stakeholder actions, words, and decisions. The program manager should maintain focus on the primary stakeholders to prevent the job of stakeholder engagement from being all-consuming at the detriment of other critical aspects of managing a program.

Stakeholder engagement work will test the courage of program managers who need to be brave and bold when faced with building relationships with stakeholders who are not a fan of the program, or may be professionally threatened by the outcome of the program. Do not follow the human tendency to avoid these stakeholders. Rather, seek them out and most importantly, listen to what they have to say. Only by listening can a program manager begin to find middle ground to use as a means to positively influence.

Use the stakeholder management techniques described above to increase your political acumen, and to become politically astute. This means understanding and being sensitive to the concerns, interests, and personal agendas of the powerful stakeholders, and using the information learned from them to counter the effects of political maneuvering on your program.

RISK MANAGEMENT

For many companies, establishing industry leadership means assuming a higher level of risk due to the uncertainties associated with navigating unchartered territories. Developing new capabilities in today's environment is risky business by nature, especially if a company wants to establish or maintain a leadership position. However, risk taking does not mean taking chances. It involves understanding the risk/reward ratio, then managing the risks associated with a program.[8]

By understanding and containing the uncertainties on a program, the program manager is able to manage in a proactive manner. Without good risk management practices, the program manager will be forced into crisis management activities as problem after problem presents itself, forcing the program manager to constantly react to the problem of the day (or hour). Risk management is a preventive practice that allows the program manager to identify potential problems *before* they occur and put corrective action in place to avoid or lessen the impact of the risk. Ultimately, this behavior allows the program team to accelerate through the program cycle at a much faster pace.

Understanding the level of risk associated with a program is crucial to the program manager for several reasons. First, by knowing the level of risk associated with a program, a program manager will have an understanding of the amount of schedule and budget reserve needed to protect the program from uncertainty. Second, risk management is a focusing mechanism for the program manager. It provides guidance as to where critical program resources are needed—the highest risk events require adequate resources to avoid or mitigate them. Finally, good risk management practices enable informed risk-based decision making. Having knowledge of the potential downside or risk of a particular decision, as well as the facts driving the decision, improves the decision process by allowing the program manager and team to weigh potential alternatives, or tradeoffs, to optimize the reward/risk ratio.

Program-Level Risk

Managing risk on a program involves many of the same tactics as managing risk on a project. The difference, however, begins with the need to view risk from an expanded perspective. At the project level, project managers on a program are concerned with risk events that will prevent successful execution of their project. Since program managers are responsible for the delivery of business benefits, risk management at the program level is concerned with the factors that may prevent full realization of those business benefits.

Figure 7.3 illustrates the two dimensions of risk management that need to be performed on a program.

The figure shows a small program consisting of three projects. As an example, the program could consist of a software project, a hardware project, and a manufacturing project. Each of the project managers are responsible for identifying, analyzing, tracking, and responding to all risks that may prevent his or her project team from delivering their component of the solution on time, within budget, and with the features and quality levels specified. Each project, therefore, will have its own specific set of risks to manage based on specialty function(s).

The program manager needs to take a holistic perspective of risk, which includes risks from each of the projects as well as risk from outside of the project, such as risks associated with the business environment and organizational factors.

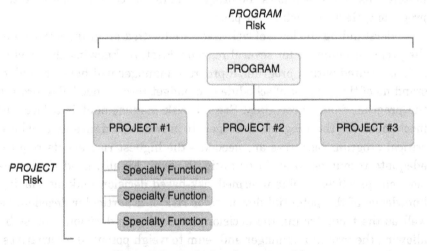

Figure 7.3 Project and program risk.

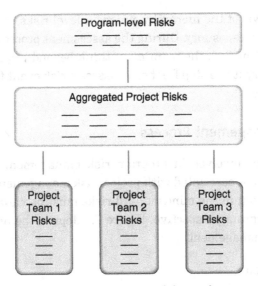

Figure 7.4 Program risk hierarchy.

This does not mean, however, that *all* risks get elevated to the program level and become the responsibility of the program manager. The key is to filter the collective risks to delineate program-level risks from project-level risks, and then divide responsibility for managing the risks accordingly. This activity creates a risk management hierarchy that is present on programs and not normally found on projects (see Figure 7.4).

The hierarchy of risk on a program consists of three levels: 1) project-specific risk, 2) an aggregation and collection of all project risks, and 3) program-level risks.

At the first level, each project team identifies the risk events specific to creating the outcomes and deliverables for their project. Each project team will have a registry of risks for which they are responsible for managing and for keeping up to date.

At the second level, the risk events for each of the projects on the program are collected in a joint-project risk repository. The joint-project risk repository is an amalgamation of all risks identified on the program. The IPT core team has the responsibility of keeping this risk registry current. As a word of caution to the program manager, know that the joint project risk inventory can potentially be very large and unmanageable. It should only serve two purposes: 1) to be the central database of risk events on a program, and 2) to distinguish risks that have both project and program-level impact.

The third level of the hierarchy, program-level risks, is a subset of the joint-project risk repository. During the assessment process for each of the project risks identified, the team must think beyond project-specific risk impact and carry it one step further to assess a risk event from a program perspective as well.

The Risk Management Process

The major steps involved in program risk management are similar in nature to the steps associated with project risk management. The primary difference is again in the mindset—all risks must be viewed from *both* a project and program perspective. Figure 7.5 depicts the primary steps of program risk management.

Risk Identification

The first step in the risk management process is to identify all of the events that could possibly affect the success of the program. Although risk identification is the first step in the risk management process, it is not a one-time event. Risk identification is an iterative process that occurs throughout the life of a program.

Risk identification is most effective when a bottom-up approach is applied. Specifically, each project team within the program should be responsible for identifying risks to the success of their project. Some program teams first begin by identifying the categories of risk, such as technology risk, market risk, business risk, and human risk, and then use brainstorming and other problem identifying techniques to identify all potential risk events within each category.

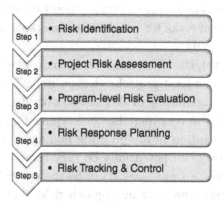

Figure 7.5   Program risk management process flow.

The key element of this step is to attempt to identify *all* potential risks. If risk identification is done well, it can be overwhelming, especially early in a program when the number of uncertainties is at the highest. Remember that the goal of risk identification is to flush out as many potential risks as possible to get them on the table for discussion.[9]

Project Risk Assessment

Not all risk events require action. The risk assessment step is needed to sift through all of the risk events identified in the previous step to identify those that pose the most serious threat to the success of each project. Scenario analysis is one of the most common methods for assessing risk events.[10] Scenario analysis involves analyzing each risk event in terms of the outcome, weighing the severity of the impact of the outcome, evaluating the probability of risk events occurring, and understanding when the risk event may occur.

An effective technique to assess the impact of a risk event is to write an "if/then" statement. The "if" is the identified risk event, and the "then" is the impact of the event. For example: *If* the user interface designers are not assigned to the program by July 1, *then* the capability go-live date will be delayed two weeks. The *impact* is a delay in the go-live date, and the *severity* of the impact is two weeks.

When assessing project risk, the project manager is responsible for evaluating the impact of each risk event on the program as a whole. This responsibility lies with the project manager as he or she represents the project at the program level and possesses a more holistic view of the goals and activities for the program and business. The risk event in the example above has the potential for being a program-level risk since it affects the go-live date.

Program Risk Evaluation

Project risks that could affect the business benefits of the program or that could affect more than one project are better managed at the program level where higher perspective and authority will increase the quality of risk responses.

Each risk identified at the project level should be entered into the collective project risk registry for review and evaluation by the program core team. Each project manager will have performed an initial assessment of their project risks and should be prepared to make a recommendation on whether a risk is contained within his or her project, or should be considered a program-level risk.

Evaluating risk from a program perspective involves asking two key questions:

1. Does the risk event have the potential to affect the business success factors of the program?
2. Does the risk event affect more than one project on the program?

A "yes" answer to either of these questions identifies the risk event as a *potential* program-level risk, requiring it to be evaluated by the IPT core team.

If a risk event has the probability of impacting one or more of the program objectives, then it has the potential to affect the *business results* driving the program and must be managed at the program level.

Likewise, if a risk event affects multiple projects on a program, it must be elevated to the program level where the cross-project interdependencies are monitored and managed. The result is a prioritized list of program risks that the IPT core team can then manage.

We recommend that a program team begin with a qualitative approach to risk assessment, at least for the first iteration of analysis. By this we mean assessing whether the severity of impact and the probability of occurrence is high, medium, or low for each of the risks identified. This gross analysis will accomplish two important things. First, it will quickly prioritize the risks so the highest risks can be identified for immediate action. Second, it gives the program manager and team an understanding of the overall risk level of the program.

If additional analysis is needed, more quantitative methods of risk assessment can be utilized on the next iteration. For the more sophisticated programs of medium and large size, the Monte Carlo analysis technique for quantitative analysis can be applied.[11] The important thing to consider, when engaged at this level of analysis, is to ensure that the value of the quantitative information gained is greater than the cost of obtaining the information. Make sure the results are worth the resources (time, personnel effort, and energy) invested.

Risk Response Planning

At this point in the process, the IPT core team has a relatively short list of program-level risks identified, assessed, and prioritized. The next step in the process is to decide which risk events need action plans. The most common risk response options available to a program team are risk acceptance, avoidance, mitigation, and transference.

Risk acceptance: This involves accepting the impact of a risk event when it occurs, but fully understanding the severity of the impact prior to its occurrence. Risk acceptance is not a common response option in low-risk tolerant companies and industries, but it is relatively common in high-risk tolerant companies and industries.

No risk mitigation or avoidance plans are needed for risk acceptance, and no changes to the program plan are required. However, a contingency plan is needed to identify alternative courses of action if the risk event occurs.

Risk avoidance: Risk avoidance involves changing the program plan to avoid the potential cause of an identified risk event, therefore removing the probability that the particular risk event will occur. For example, a program team may recommend removal of a particular new technology because the risk of the technology is too high and may adversely affect the program's success. Risk avoidance requires a risk-response plan, as well as appropriate changes to the overall program plan and relevant project plans.

Risk mitigation: There are two basic strategies for mitigating a risk event: 1) reducing the probability that the risk event will occur, and 2) reducing the impact that the risk event will have on the program if it occurs. Risk mitigation is a powerful technique because it targets the root cause of the risk event.[12] It requires a risk-response plan and may also require changes to the program plan.

Risk transference: If a program team recognizes that it lacks the expertise or the ability to respond to a risk event internally, they may choose to transfer the risk response to another party. This, however, neither removes the risk from the program, nor does it remove the responsibility for the risk from the program manager. It simply transfers the response to the risk. Warranties and insurance are a form of risk transference, as is outsourcing of knowledge work.

Risk Tracking and Control

The last step in the risk management process involves monitoring the trigger events associated with the program risks, identifying new risks, and executing risk response plans or contingency plans when risk events occur. The program manager and the IPT core team need to track and control the program risks with the same diligence that they track and control the program schedule and finances. They also need to keep an open mind that new risk events will come to light throughout the duration of the program that will need to be assessed and potentially responded to.

Risk Management Practices

Risk management should not be viewed as a one-time event. It's not uncommon practice for a program team to present the risks associated with a program at each decision-checkpoint meeting, then set the information aside until the next decision. Practicing risk identification without practicing risk management is a fatal flaw. Risk management must be practiced consistently throughout the life of a program.

To ensure successful implementation of risk management practices on a program, a safe environment should be created for the team to discuss potential problems, an understanding should be established that risk management is not an exact science, the risk-tolerance levels of program stakeholders should be considered, and risk-based buffers to safeguard against the risks that cannot be resolved or anticipated should be utilized.

Don't Shoot the Messenger

For the risk management process to be effective, the program manager needs to establish an environment in which the team members feel comfortable raising potential problems and concerns. If the program environment is such that bad news and mistakes are punished rather than tolerated, team members will be reluctant to speak freely. This will choke the risk management process.

Risk Management Is Not an Exact Science

When we say risk management is not an exact science, we're referring to the fact that the program manager is always working with subjective risk information. After all, we're trying to predict what *may* happen in the future, not analyzing what *did* happen in the past. Therefore, two things come into play. First, it's not possible to identify and manage every risk. There will always be problems that have to be contended with (problems are risks that weren't identified or managed). Second, don't get lost in trying to *quantify* every risk. This can be contentious because many people have created quantification models and algorithms to help with managing risk. These models and algorithms are great tools for people in some industries, but in most industries there are so many potential risk events that a program manager could be consumed by risk modeling. For these program managers, learn to trust your experience and instincts.

Consider the Concept of Risk Tolerance

Risk tolerance is the level of risk an organization is willing to accept to achieve its business goals. Every organization and every individual within an organization has a different level of risk tolerance, with corporate culture and values being a primary influencer of acceptable tolerance levels. Figure 7.6 illustrates risk tolerance as a continuum, from complete risk avoidance on one end to complete disregard of the consequences of risk on the other.

For example, companies whose products involve human lives (automotive, aerospace, medical) have a low tolerance for risk due to customer requirements and product use conditions. For these companies, low risk tolerance means many risk avoidance steps are included in the program plan. In contrast, in some industries (application software, home electronics) risk taking is necessary to attain and maintain a competitive advantage. As a result, many risk events are accepted, program plans are aggressive, and schedule and budget contingencies are used to allow for some level of risk impact.

So, how does this affect the program manager? First, the program manager must understand the individual risk tolerance of primary stakeholders involved with the program, especially the sponsor, functional managers, project managers, and the governance board. Second, understanding tolerance levels provides excellent guidance for how much risk a program manager can assume, and what response actions (avoidance, acceptance, mitigation, or transference) are accepted as preferred options within the organization.

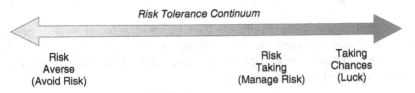

Figure 7.6   The risk tolerance continuum.

Unfortunately, there is no mathematical formula, no statistical calculation, and no model to help manage the various levels of risk tolerance the program manager will encounter—that's why we refer to this as the "art" of risk management. Experience, understanding the corporate culture, and developing personal interaction with the people on and surrounding the program are the best methods for learning to manage organizational risk tolerance.

Utilize Risk-Based Contingency Buffers

Because programs operate in a less than perfect world, savvy program managers use the information they gain from risk management practices to safeguard their programs against the suite of risks that will occur regardless of the best efforts to avoid them, and the suite of risks that simply were not able to be identified in a proactive manner. One of the most effective ways to safeguard a program is by applying contingency buffers to the program timeline and program budget.

Based upon the risk analysis step, the program manager can estimate the buffer amount he or she needs to include to increase the probability of program success. There are multiple ways to estimate the buffer amount. Three of the most common and effective approaches are: 1) utilize historical schedule and budget variance data from previous programs, 2) employ critical chain scheduling tools (schedule only),[13] and 3) use simple statistical analysis techniques (see "Calculating Schedule Buffer").

The job of the program manager is to determine the amount of risk the organization is willing to assume, calculate the amount of risk contingency needed, and then add it to the master program schedule and program budget. It is important to note that the management style and philosophy of the senior leaders for various companies all have different risk tolerance thresholds and therefore it is recommended that the program manager discuss risk tolerance and risk buffer strategies with their management to ensure agreement on the risk strategy for the program.

Calculating Schedule Buffer

One method of estimating schedule buffer that we have found accurate and useful is the simple statistical analysis technique known as three-point estimation. The following formula for a weighted average-per-beta distribution can be used to calculate expected program duration:[14]

$$Duration = \frac{o + 4m + p}{6}$$

Where:

"o" is the OPTIMISTIC duration

"m" is the MOST LIKELY duration

"p" is the PESSIMISTIC duration

The initial time duration generated from a program mapping exercise is commonly used to determine the optimistic duration. For the pessimistic duration, the team can use the risk analysis data that has the most significant schedule impact and add it to the optimistic duration. Historical variance averages for similar programs can be added to the optimistic duration to yield the most likely duration.

For example, suppose we have a program that has a duration of 16 months as determined by program mapping—this is the optimistic time. Historical data shows similar programs have been late by an average of 10 percent. Ten percent of 16 months (1.6 months) is added to the optimistic duration to give a most likely duration of 17.6 months. The risk analysis shows a maximum schedule impact of three months for one of the high risk events identified. This is added to the optimistic duration to give a pessimistic duration of 19 months. The times and estimated schedule duration are as follows:

Optimistic time = 16 months

Most likely time = 17.6 months

Pessimistic time = 19 months

$$Estimated\ duration = 0 + 4m + p = \frac{16 + 4\,(17.6) + 19}{6} = 17.5\ months$$

The good news is that the calculation above gives a solid statistical estimate for schedule buffer of 1.5 months. The bad news is that it will only raise the probability of schedule success to about 68 percent. Most program managers are not comfortable dealing with a 32 percent chance of failure. To increase the probability of success, a standard deviation can be added to the expected program duration calculation. Where one standard deviation is defined as follows:

$$S = \frac{p - o}{6}$$

For the example above, the standard deviation is:

$$S = \frac{19 - 16}{6} = 0.5\ months$$

By adding one standard deviation (half of one month) to the estimated time calculation above, the probability of schedule success increases to approximately 80 percent, two standard deviations added (one month) equates to 95 percent probability, and three standard deviations (one and one half months) give the program manager a 99 percent probability of schedule success. We recommend not going above a 95 percent confidence level.

If two standard deviations (one month) are added to the estimated schedule above, the total duration of the program is 18.5 months. This calculation tells the program manager that by adding 2.5 months of schedule buffer, the probability of schedule success is now at 95 percent. By adding a little over 15 percent risk contingency to the best-case schedule, the probability of success has nearly doubled.

FINANCIAL MANAGEMENT

Since program management is part of the business engine of an organization, good financial management practices are a necessity for consistent success in managing an organization's programs. This involves setting clear financial targets, cost estimation and control, financial feasibility analysis, cash flow management, and budgetary management.

Develop Financial Estimates

The first step in financial management is to develop budgetary-level financial estimates as part of the definition and planning activities of a program. This may include program budget estimates for use in the program business case, return on investment estimates, financial break-even estimates, and estimated average selling prices or capability cost. This activity requires a strong partnership with a financial representative for the program who understands the intricacies of the financial department of the company. In fact, many firms ensure that a financial department representative is assigned to more business critical programs.

Determine Financial Feasibility

As part of the program business case, the program manager is responsible for a robust financial feasibility assessment. The feasibility assessment tests the viability of the program by evaluating the financial estimates developed in the previous step to determine if the program is a viable investment option for the business. This analysis should answer the following questions:

- How much will the program cost to implement?
- How much will this program contribute to the company bottom line?
- Is the program worth investing in?

Develop Project Budgets

Once the financial feasibility of a program is determined and approved, the next step is to engage in detailed cost estimation. The detailed cost estimation activities occur at the project level and are "rolled up" and managed at a summary level by the program manager. Each project manager uses traditional project-cost estimation techniques and information contained in his or her project work breakdown structure, detailed schedule, and staffing plan to estimate the resource costs for the project. In addition, any fixed costs such as equipment and subcontractor support should be included in each detailed project budget.

Integrate the Project Budgets

With the detailed project budgets complete, the program manager can integrate them into a program budget that represents the total investment in the program. Any costs outside of the project budgets, like the cost of the program manager's salary or subcontractor cost, are added to the program budget in this step. The program budget should include all variable costs, fixed costs, and overhead charges that the program will incur. The budget should encompass the cost to develop a capability and also the cost of program operations following the delivery of a capability.

Add Budget Reserve

Using the risk assessment data generated, the program manager can estimate the amount of budget reserve he or she needs to include to increase the probability of program financial success. By taking an average financial exposure—for the top ten risks, for example—the program manager can determine the amount of budget risk he or she needs to add. Another commonly used technique is to use a standard percentage of the overall budget; a good rule of thumb is 5 to 10 percent. Once estimated, the program manager should add the risk contingency to the budget to develop a high-probability program budget that he or she can put in front of senior management for negotiation and approval. It is recommended that the program manager discuss the use of this strategy and amount of the budget reserve with the program sponsor and senior leaders of the organization.

Negotiate Final Targets

In this step, the program manager negotiates the financial elements of the program with the senior sponsor or the customer. Financial targets such as the ROI, profit margin, manufacturing costs, and time-to-breakeven can all become negotiable items as long as they still comply with the business goals of a program. In addition, the program budget itself, especially the budget reserve, is commonly negotiated. It helps tremendously if the program manager comes to the negotiating table armed with the data supporting the financial elements of the program to focus the negotiation on the data, not the overall targets. At the end of the negotiation, program managers have a set of program financials that senior management will hold them accountable for achieving.

Manage and Control Program Finances

The program team now has the financial information needed to manage the remainder of the program. Managing program finances involves tracking actual progress against budget and comparing it to planned progress. This should be performed in a two-tiered approach. The project managers should manage the financial details within their projects, and the program manager should manage the finances at the summary level.

When a variance between planned versus actual financial performance is detected, or when issues and changes are encountered, the variance must be effectively controlled to ensure the financial viability of the program. The specific control techniques that are employed will depend upon the type of financial variance encountered. For example, if the bill of material cost for a product exceeds the targeted amount, the team can look for lower cost part substitutes or recommend removal of one or more features to eliminate material cost. Managed consumption of the budget reserve is also a viable option for managing budgetary variances.

The program manager also has to monitor and control the cash flow on the program, especially in smaller or cash constrained companies. Once a program enters the implementation phase, it can become a large financial drain on the company. Jeff Singleton, a former Boeing program manager, explained that on a program consisting of 100 people (large by some standards, low by others), his program expenditures exceeded $120,000 per day to pay for the work of his program team. That's $600,000 per week!

Effective cash flow management is necessary for the program finances to remain solvent.

Financial Management Practices

To ensure successful implementation of the financial aspects of a program, special consideration should be paid to the financial competency of the program manager and utilizing an effective budget estimation process.

Financial Competency

As a business leader within a company, the program manager must possess strong financial competency (Chapter 11). In the role of the business manager proxy, the program manager assumes the financial responsibility of a program from the business manager. This means that he or she needs to have the financial acumen to carry out this responsibility. Short of getting a degree in finance, the program manager should look for education and training opportunities to bolster his or her ability to understand, manage, and speak to the financial aspects of managing a program.

Budgeting Methods

We have witnessed multiple methods used to develop detailed budget estimates, most of which have been successful. The most common are bottom-up estimation, top-down estimation, and iterative estimation. Any of the three will work, as long as both the program manager and his or her senior management can come to an agreement on the financial targets. From the program manager's perspective, it is our experience that the iterative approach to budget estimation yields the best opportunity for deriving a budget that has the maximum buy-in from both the program and senior management teams.

CHANGE MANAGEMENT

Rest assured, change *will* occur on a program. This is primarily due to the fact that programs exist within dynamic and ambiguous environments— the very reason why program management is a preferred management approach for these environments.

Effective program change management practices begin with a fundamental change in mindset. Most humans, even those populating a program team, are not fond of change, especially if it affects them personally. As a result, they tend to resist change.[15] However, if change is a near certainty for a program, this resistance to change must be dealt with through establishment of a pro-change mindset by viewing change as an opportunity to improve. The program manager must be the advocate for controlled changed on a program.

This underscores the basic difference between change management practices for projects versus programs. On projects, change is viewed as a deviation from the project baseline that needs to be corrected in order to realign to the existing baseline. On programs, change is viewed as a result of something learned and therefore an opportunity to improve the business benefits delivered. Thus, an opportunity exists to establish an *improved* program baseline. This change in mindset is necessary due to the uncertainty that normally surrounds a program.

Change management practices allow for uncertainly and ambiguity in the program environment, while at the same time keeping change bounded by the established business success criteria. Program change management is about enabling change within a predefined, desired business end state.

Change Management Elements

For consistency in practice, it is important to have a documented change management process. The process is normally part of the program-level governance system and does not have to be complicated to be effective. In fact, effort should be expended to simplify the process as much as possible.

Figure 7.7 illustrates an example of a change management process that is used by a major automobile manufacturer. This can be used as a starting point for developing a customized version for any organization.

Submitting Change Proposals

The first step in the change management process is for the person requesting the change to submit a written change request proposal. This proposal should include a detailed description of the change requested, a justification for the change, and the benefits to the program for implementation of the change.

The justification and benefits are critical pieces of information that the program manager should require for any change request. Cursory statements such as "the customer has requested the change" provide good

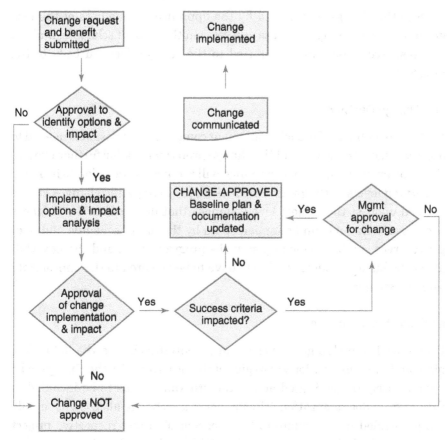

Figure 7.7 Change management process.

reason to *consider* a change, but are not good enough to *approve* change. The change request proposal should present a compelling argument for the program team to change its current course of action.

Assessing a Change Proposal

Change assessment involves evaluation of why the change is necessary, if it is in alignment with the business success factors, and what it will cost and require for implementation. The assessment culminates in a change benefits/cost/impact analysis.

Change proposal assessment takes resource time and effort away from members of the program team that would ordinarily be spent developing their project deliverables. By filtering poorly justified or non-value-added changes, the change review board prevents wasted effort on the part of program resources.

Once the change is assessed by the appropriate members of the program team, the change benefits and costs in the form of schedule, budget, and resource impact are submitted to the change decision body of the program.

The Change Decision

With the benefit/cost/impact assessment completed, a decision is needed to approve, disapprove, or send the change proposal back for further evaluation. The level of impact of the change will require the decision to be made at varying levels of the program. Tactical, project-specific changes can be decided upon at the project level; changes that don't affect the business success criteria and can be implemented by the use of schedule and budget reserves can be decided upon at the program level; and changes that affect the business success criteria have to be elevated to the appropriate senior manager.

Implementing a Change

If approved, the change proposal can potentially change the definition, plan, and implementation strategies of the program. For the change to be appropriately disseminated across the program, it has to be included in the program documentation. The impact of the change also has to be fully comprehended in an update to the program plan and respective project plans. This includes new or modified deliverables and tasks, changes to the timeline and schedule, and possibly changes to the program budget.

Communicating a Change

The last step in the change management process is to broadly communicate the change and the impact of the change to all affected and interested program stakeholders. Communication should include the change description, impact to the program, and the benefit of implementing the change. This follows our philosophy of no surprises to management (and customers).

Change Management Practices

Change management should begin on a program as soon as the requirements are identified. This can be as early as the program definition if it makes sense to manage the high-level requirements driving the concept and business-case development efforts. However, change management processes and protocols must be in place once the detailed requirements begin to be documented. If the requirements continue to change without being adequately managed and documented, the outcomes from program execution will not accurately reflect the program requirements. This situation results in the need for reconciliation between the requirements and work outcomes later in the program, when it is much more costly and painful. It is better to begin managing the changes to the program requirements *before* the program plan is fully developed, communicated, and committed.

During program execution, changes have to be tightly managed. Even the smallest changes can have significant impact due to the potential high cost of rework and risk of performance errors once the output of a program is operational.

Change Management Hierarchy

Best practice organizations often implement a change management hierarchy for their programs. At the lower level, change is managed within each of the projects, and is focused on tactical changes specific to the particular project discipline. A change review board is typically used and consists of a subset of the project team leaders and is led by the project manager. To be effective, the program manager must empower the project managers to make the appropriate change decisions.

At the higher level, cross-project and business specific change is managed by the program manager and a subset of the IPT core team.

Change Management Protocols

Change management protocols define how changes are to be addressed once the program is underway and should be clearly communicated by the program manager.[16] The purpose of protocols is to ensure the change management process functions as efficiently as possible. The easy part of establishing an effective change management process on a program is defining the process; the hard part is changing the behavior of the people expected to follow the process. Change management protocols are meant to help with the behavioral challenges that will be encountered.

Example protocols include the following:

- All proposed changes must be made in writing and include justification for the change and benefit to the program and business.

- The change management process must not be sidestepped by escalation to a senior manager.
- All change decision members are required to attend all change management meetings or send an empowered delegate.
- Any change impacting the business success factors must be elevated to a designated senior manager for approval.

Fast-Track Change Requests

To ensure successful implementation of a change management process on a program, special consideration should be given to establishing a fast-track process for time-sensitive decisions. In a fast-track process, all change requests must be reviewed and evaluated in the shortest time possible to prevent unneeded overhead burden on the program. This is normally instituted through an email system that requires a minimal set of decision makers and limited time to review, analyze, and recommend a change decision. For any fast-track change approval practice to remain effective, it needs to be maintained as the exception, rather than the norm.

PROGRAM-LEVEL GOVERNANCE

Program-level governance is a subset of organizational governance, and is performed within the context of the overall organization.[17] At the highest level, a firm's board of directors governs an organization's strategic direction and investment decisions. The senior executive team governs the direction of various business and operational units within the firm. The program governance board provides oversight of the various programs in flight within the organization, and the program manager establishes the program-level governance for a specific program.

The program and project management standards bodies have provided a variety of governance models. In practice, we have found that the models seldom provide a perfect solution for an organization. The reality is that an organization must first understand where it sits with respect to program management maturity, and then tailor a governance system based upon their needs. To demonstrate, we use the program management continuum first introduced in Chapter 1, and modified as illustrated in Figure 7.8.

If an organization is on the far left of the continuum, or administration-focused, it is likely that no formalized program governance model exists, or

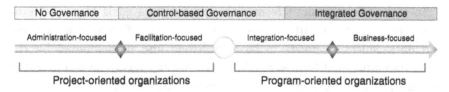

Figure 7.8   Program-level governance continuum.

is needed. For organizations that operate in the middle of the continuum, it is common to see a traditional controlling approach to program-level governance. This is because a single-project approach with traditional project management methods remains the primary philosophy and influence. The collection, analysis, and reporting of project performance data, with the intent to measure conformance, is the primary approach.

For program-oriented organizations, those operating on the right side of the continuum, an integrated approach to program-level governance is necessary.[18] The integrated approach recognizes the collaborative relationship between the project teams, the program team, and the business team of an enterprise.

An integrated approach to program-level governance incorporates many elements, but is primarily centered on three factors: 1) setting and maintaining strategic alignment, 2) reviewing progress, and 3) making decisions that affect program strategy.

Setting and Maintaining Strategic Intent

Program-level governance begins with ensuring that a program and its constituent projects are aligned to the strategic goals of an organization. In the early stages of a program (program definition and investment), appropriate due diligence should be applied to create alignment between the strategic goals driving the need for a program, the business results the program will deliver, and the program objectives. Program-level governance then ensures that there is alignment between the outcomes and deliverables for each of the projects to the program objectives. We detail the process of benefits translation and alignment in Chapters 3 and 6 respectively.

Setting strategic alignment is a critical first step. Maintaining alignment as a program progresses through work cycles and time is equally important and continual focus and priority on the part of the program manager and the IPT core team is necessary.

To maintain alignment from strategic business goals to project outcomes throughout the life of a program, a system of measures, metrics, and data collection needs to be established and implemented by the program manager (Chapter 8). This system then has to be incorporated into the program work cycle and continually used to ensure alignment is maintained and adjusted if and when it becomes necessary.

Reviewing Program Progress

A program manager spends a lot of his or her time communicating the status of the program to various stakeholders. This reporting takes the form of both formal and informal communications, from hallway conversations to formal program reviews and decision checkpoint meetings with senior management and other key stakeholders. Regardless of whether the status is formal or informal in nature, the message should remain consistent.

Consistent communication of program progress is a critical element of cross-project collaboration. Any significant deviations or changes to the program must be communicated to the project teams by the program manager. In like fashion, any changes that have occurred on any of the projects must be communicated to the program manager and other project teams by the project managers. Late or ineffective communication can quickly result in rework, delays, and added cost to a program.

So, how does the program manager know he or she is reporting a consistent, comprehensive, and accurate message about a program? The key is in consistent, comprehensive, and accurate collection of status from the project teams and other integrated core team members.

IPT Project Reviews

The IPT core team should review detailed status for each of the projects on a regular basis. For most programs, the primary agenda for each core team meeting is project status. Reporting project status facilitates the flow of information within the program and importantly, intentionally creates a means for conversation about expectations and integration points among the project teams.

The program manager needs to be specific on what elements of the projects he or she wants reported, what metrics to use, and what level of detail to include to gain consistency in reporting across the project teams.

In practice, this will take some work and time in coaching the core team members, especially with a new team.

An effective tool for establishing cross-project reporting consistency and concise communication of project status during the internal program status reviews is the project indicator. An indicator is a brief, one- to two-page document that shows pertinent project status (Chapter 9).

By working with the project teams in developing a consistent format for the indicators, and a consistent set of metrics to report, the program manager will receive a comprehensive and concise report of project status that he or she can use to develop a formal or informal program report.

Program Reviews

The program review is an organizational meeting in which program status is reviewed by the program governance board or the executive sponsor. Each program manager is required to present the status for the program he or she manages. Even though the program review is primarily a status meeting, it can also be used as a decision-making forum for factors that affect the program objectives.

For consistency in reporting format and message, it is most effective to have an established program indicator for use by all program managers. Much like the project indicator, the program status indicator is a one- or two-page summary brief that shows pertinent program status (Chapter 9). Much of the information contained in the indicator is derived from the most recent project indicators and presented in summary format.

In addition to the program status indicators, the program strike zone and program dashboard tools are normally used in the program review to communicate status toward achievement of the program success factors and the business goals (Chapter 9). The objective for the program manager in the review is to communicate both operational and strategic status of the program at the executive level and to use the forum to gain executive help if needed.

Keep in mind that the program review serves dual purposes. It provides the data and information necessary to ensure that a program and its constituent projects are progressing as anticipated, and it provides the data and information necessary to enable the senior executive or program governance body to make good decisions with respect to business value and strategic direction.

Making Effective Decisions

Due to the complexity of most programs, program teams need to draw upon a wide array of information to make effective decisions. This means all disciplines and projects must be involved in generating information for and participate in the decision-making process. Because of the cross-discipline and cross-project nature of programs, information must be presented in a manner that all parties can understand and be able to use in the decision process. It is no longer the case that engineering utilizes technology-only data, or marketing utilizes customer-only data, or finance utilizes accounting-only data to drive program-related decisions. Discipline-specific data and information must be combined with cross-discipline information for program-level decisions.

An effective program decision is described as one that has the following characteristics:

- Not reached by a single individual
- Based upon input provided by each team member involved in the decision process
- A sound solution to the problem
- Aligned with the program objectives and business success criteria

Program teams are charged with making three types of decisions: 1) strategic decisions, 2) tactical decisions, and 3) operational decisions. Strategic decisions may include such things as which technologies to include in a design, what product cost is needed to maintain competitive leadership, and what key partnerships or alliances in which to engage. Tactical program decisions may include which features to incorporate, how many prototypes to build, and how to manage changes to the program schedule. Operational program decisions may include how to coordinate factory scheduling, which common cross-project processes and tools to employ, and how to communicate progress to program stakeholders.

Effective program decisions require well-coordinated and communicated interplay between the project teams and empowerment on the part of organization's management team to make the strategic, tactical, and operational decisions necessary. By empowering the program team to make decisions, decision making becomes quicker and more effective. If a program team is required to rely on functional or line managers for decisions, these managers will quickly become a bottleneck in the program's implementation, thus, reducing efficiency and effectiveness.[19]

Program team empowerment means giving the program manager and his or her project managers the responsibility and authority to make decisions. In the book *The Power of Product Platforms*, Marc Meyer and Alvin Lehnerd state, "There is no organizational sin more demoralizing to teams than lack of empowerment."[20] The program team must have confidence that their empowerment will not be undercut by the firm's management. If such action occurs, it must be immediately eradicated by the firm's senior management (see "The Battle of Powerville").

Just as the program manager must be empowered by senior management, the program team must also be empowered by the program manager to make key decisions. As Rich Nardizzi, a senior program manager from a major aerospace company told us, "Empowerment is something that builds over time through gaining the confidence of the leader." He also pointed out that there needs to be empowerment boundaries because, as he put it, "One big screw-up on the part of someone wipes out ten attaboys."

The Battle of Powerville

Rollercoaster History: Founded in the early 1900s, it took Miner Truck Company more than 60 years to reach the number one position in the mining vehicle industry—a position they would hold for nearly 30 years. However, in the early 2000s, things began to change. Annual sales began to drop, financials were in the red for several years, and the company dropped to number three in the industry. Additionally, it took the company 16 months to develop a new product, while competitors needed only 8 months. As a result, Miner Truck became a takeover target, and soon a Swedish multinational acquired the company—ending the family-owned business of nearly 90 years.

Hopeful Future: Under new senior leadership from the Swedish parent company, things again began to look rosy during the next five years. Sales grew 200 percent, annual ROI increased to a respectable 14 percent, and the product development cycle time became shorter. For this to happen, some fundamental changes were made. The old functional structure was abandoned for a matrix structure and product development teams became cross-functional. The teams were led by a program manager and were empowered to make major program decisions. The original functional vice presidents (VPs) remained, but their decision-making power was transferred to the team. The first pilot program achieved the time-to-market goal of eight months, which was in line with the company's competitors.

Failed Future: The leaders of the functional groups had the power to nominate the team members that would represent the functions on the programs. Despite new team charters that empowered the team to handle decisions within their scope, the VPs saw this as an erosion of their power within the company. They therefore ordered their representatives on the team not to make any decisions without consulting them first. At first the team members resisted, but later acquiesced over fear of negative effects to their careers. This turned the program team from a fully empowered program team to a powerless coordinated project team, and the VPs into the primary decision makers for the program.

It took the senior management team of the Swedish parent nearly a year to figure out why schedule performance on new product development programs was slipping dramatically. One month later, all the VPs and one program manager were let go.

Lessons learned from this story include:

- Major organizational changes are not accomplished without shifts in the power structure.
- Shifts in the power structure can lead to power struggles on programs if not overseen by executive management.
- Organizational interest is better served once the power struggles are resolved in alignment with the organizational goals.

ENDNOTES

1. Program Management Academy LinkedIn Group, www.linkedin.com/groups?gid=118003&trk=my_groups-b-grp-v.
2. Project Management Institute. *The Standard for Program Management,* 2d ed. Newton Square, PA: 2008.
3. CIO Executive Council. "The State of the CIO '08." *CXO Media,* 2008.
4. Pinto, Jeffrey K. *Power and Politics in Project Management.* Newtown Square, PA: Project Management Institute Publishing, 1998, p. 27.
5. Roeder, Tres. *Managing Project Stakeholders: Building a Foundation to Achieve Project Goals.* Hoboken, N.J.: John Wiley & Sons, 2013.
6. Bourne, Lynda. *Stakeholder Relationship Management.* Burlington, VT.: Gower Publishing, 2009.
7. Pinto, *Power and Politics.*
8. Martinelli, Russ, and Jim Waddell. "Managing Program Risk." *Project Management World Today,* September–October 2004.
9. Smith, Preston G., and Guy M. Merritt. *Proactive Risk Management.* New York: Productivity Press, 2002.
10. Frame, J. Davidson. *The New Project Management.* San Francisco, CA: Jossey-Bass Publishing, 1994.
11. Milosevic, Dragan Z. *Project Management Toolbox: Tools and Techniques for the Practicing Project Manager.* Hoboken, N.J.: John Wiley & Sons, 2003.

12. Smith and Merritt, *Proactive Risk Management*.
13. Goldratt, Eliyahu M. *Critical Chain*. North River, MA: Great Barrington, 1997.
14. Keogh, Jim, Avraham Shtub, Jonathan F. Bard, and Shlomo Globerson. *Project Planning and Implementation*. Needham Heights, MA: Pearson Custom Publishing, 2011.
15. Davidson, *The New Project Management*.
16. Keogh et al., *Project Planning and Implementation*.
17. Sasghera, Paul. *Fundamentals of Effective Program Management: A Process Approach Based on the Global Standard*. Plantation, FL.: J. Ross Publishing, 2008.
18. Office of Government Commerce. *Managing Successful Programmes*. 3d ed. Norwich, UK: Office of Government Commerce, 2007.
19. McGrath, Michael E., Michael T. Anthony, and Amram R. Shapiro. *Product Development: Success Through Product and Cycle-time Excellence*. Stoneham, MA: Butterworth-Heinemann Publishers, 1992.
20. Meyer, Marc H., and Alvin P. Lehnerd. *The Power of Product Platforms*. New York: Free Press Publishers, 1997.

Chapter 8

Program Metrics

Metrics are powerful. The effective use of metrics can help senior managers understand their position relative to specific competitors as well as the overall market. They can help portfolio managers evaluate investment effectiveness as well as opportunities relative to organizational capacity. They can help program managers better understand the implication of forecasted problems as well as better articulate successes. They can also serve project teams within programs to assure alignment and performance excellence amid complex interdependencies. In short, metrics are powerful; or perhaps more precise, *for better or worse* metrics are powerful.

Metrics are part of a larger organizational construct—the performance management system—and if not used properly, they can do as much harm as good. Hence, the "for better or worse" caveat noted above.

Most organizations struggle with performance management. Common errors include measuring too much, measuring not enough, measuring the wrong things, using measures as absolutes rather than indicators, measures that are not timely, measures that are not aligned to strategic objectives, measures that create bureaucracy, measures that are not agreed to nor regularly used by senior managers, and the list goes on. While performance management is hard, it is a discriminating factor separating industry leaders from followers, those that are noted as best-in-class as compared to the rest in class. According to the research findings from Mollie Lombardi and Jayson Saba, top performing organizations use assessments to gather metrics more broadly, more frequently, and more consistently for decision making than their lower-performing competitors.[1]

This chapter provides detail that allows program managers to move toward the effective use of metrics as a powerful resource within their programs and organizations. To do so, we first want to make sense of metrics. With an understanding of metrics in place we will then explain why program management metrics are needed and explain various types of metrics and how to determine which program metrics are important based on organizational strategy.

MAKING SENSE OF METRICS

It is interesting that performance management is so hard, especially since we manage performance all the time. Each of us uses performance indicators, measures, and metrics every day and we do so relative to a baseline or set of expectations. For example, if you drive a car, you are constantly evaluating your use of the accelerator as indicated by the measure miles (or kilometers) traveled per hour, as illustrated on your dashboard, and against the baseline or expectation of allotted legal speed on the roadway. For another example, let's look at the human body. Measures of body temperature, blood pressure, body mass index, cholesterol count, and blood sugar level reflect a state of overall health. When evaluating businesses, as yet another example of measuring performance, analysts are often interested in knowing about percent of market share, return on assets, cost of goods sold, customer satisfaction, and cash flow, all of which explain organizational health relative to competitors and the overall industry. In short, we all manage performance and do so with performance indicators, measures, and metrics.

Program managers use metrics to assess program health relative to progress toward achieving business goals, value creation, and organizational benefit. In particular, using performance metrics help program managers and their sponsors understand how well a program is performing, where and why a program has problems, and tailor actions to eliminate the problems. This will, in turn, improve the program and bring it closer to its goals. Therefore, devising and employing appropriate metrics should aim to improve business results of the organization.

Program metrics not only measure the health of individual programs, but also show the effectiveness of program management-related processes, such as strategic management and portfolio management. In this manner, program management metrics are an effective means to integrate and synchronize strategic, portfolio, planning, and execution activities.

The Confusion Surrounding Metrics

While most organizations and program managers specifically, recognize the importance of metrics in general terms, the lack of use and inconsistency of use is more prevalent in practice. We have observed what Norman Fenton and Martin Neil noted in their research, which is that metrics and activities associated with metrics "have not addressed their most important requirement: to provide information to support quantitative managerial decision-making."[2] So why not? The answer is *confusion*. It is easy to get overwhelmed by the complexity of metrics, especially now with data becoming easier to obtain. Just because data are easy to obtain does not mean they are useful. We will make sense of translating data into useful information in the coming sections. For now, let's address the confusion as it pertains to nomenclature.

Much of the challenge in designing, developing, and establishing a useful performance management system is in part due to confusion in using performance terms such as objective, metric, measure, indicator, critical success factor, and more. Some authors use these performance management terms seemingly interchangeably, while others delineate them with a level of sterile technicality. Although authors may not agree, effectively establishing a performance management system in an organization requires a common language of terms and meanings. The following defines and explains key performance management system terms.

Objective: A specific target to be achieved most often associated with a goal or strategy. A major league baseball player may have, as an example, an objective to achieve a .300 batting average or hit 40 home runs and steal 40 bases. From a business perspective, an objective may be to achieve 10 percent new market penetration in each of the next three years.

Business Success Factor: Many practitioners use the term *critical success factor* (CSF). D. Robert Daniel, of McKinsey & Company, introduced the term in 1961 when his article "Management Information Crisis" appeared in *Harvard Business Review*.[3] We, however, use the term *business success factor* (BSF) so as to keep focus on business outcomes— it's all about the business. This approach is consistent with Daniel's work in that he highlighted CSFs as types of data useful in measuring and informing top management activities. Therefore, critical business success factors are organizational conditions, practices, or elements necessary for success. For example, a critical business success factor may be to acquire new customers by expanding use of social media.

Metric: A quantitative measure with an agreed-upon dimension. One inch, 132 pounds, and 4 meters are all examples of metrics. Each is specific and has an agreed-upon quantitative meaning. An inch, for example, is one-twelfth of a foot or 2.54 centimeters. From a business perspective, return on investment (ROI) of 14 percent, market penetration of 21 percent, and deliverable efficiency to schedule of 98 percent, are examples of metrics.

Measurement: A quantification process, calculation, or algorithm that yields a numeric score. For effectiveness, a measurement yields not only the score, but aligns that score back to strategy and objectives. Business leaders and investment analysts are often measuring organizational return on assets, market share, and cash flow regarding organizational performance and alignment with strategy and objectives. Again, for clarity purposes, measurement is the process of measuring metrics (the actual performance) and relating them to strategy and objectives (the performance target).

Indicator: A prescribed state(s) to notify, alarm, or warn of conditions. The use of a household smoke detector is an indicator. Another example of an indicator is the engine light on the dashboard instrument panel of a car. The indicators may go off by signaling an alarm (visually or audibly), which prompts action. From a business perspective, examples include cost or schedule overruns or underruns relative to prescribed thresholds. Early warning systems such as the program strike zone and program indicators (Chapter 9) are valuable—the earlier the warning, the better.

Rationale for Using Metrics

There is one main, often cited, and logical reason for using metrics. It is common knowledge that "what gets measured gets managed; and what gets managed gets improved." All organizations want to achieve continuous improvement. Those actively, and effectively, using performance metrics are more likely to succeed with continuous improvement. John Hauser and Gerald Katz illustrated this quite nicely in their article "Metrics: You Are What You Measure!" Here is how they explained it:

> The link is simple. If a firm measures a, b, and c, but not x, y, and z, then managers begin to pay more attention to a, b, and c. Soon those managers who do well on a, b, and c are promoted or are given more responsibilities. Increased pay and bonuses follow. Recognizing these rewards, managers start asking their employees to make decisions and take actions that improve the metrics. (Often, they don't even need to ask!) Soon the entire organization is focused on ways to improve the metrics. The firm gains core strengths in producing a, b, and c. The firm becomes what it measures.[4]

As illustrated in Figure 8.1, metrics measure performance, and that performance yields output of work effort and resource investment, which in turn determines business results. The results measure the degree to which a mission is accomplished or not.

In particular, using a performance management system helps program managers and their sponsors understand the health of a program. Metrics quantify the degree to which objectives and overall strategy are being achieved. Indicators help to pinpoint if, where, and why a program has problems, which also helps to understand root causes of problems to tailor specific actions to further mitigate or eliminate them. Therefore, when done properly, devising a performance management system and employing appropriate metrics improve the probability of program success and business benefit.

It is important to note that metrics can only tell part of the story. They cannot fully identify, explain, or predict what will happen. Most potential problems need more than one metric or source of information to properly characterize or understand an issue or problem. It is a situation analogous to assembling the pieces of a puzzle. One piece of the puzzle does not depict the image of the whole puzzle.

A useful and effective metric should be timely, accurate, relevant, objective, and presented in a useful format to aid in management and decision making. Metrics should support, not replace, good management judgment.

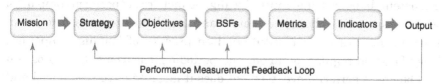

Figure 8.1 Performance management system process.

Metrics and Program Governance

Metrics play a role in support of the organization's governance processes and procedures. As pointed out in PMI's *The Standard for Program Management,* "To support the organization's ability to monitor program progress and strengthen the organization's ability to assess program status and conformance with organization's controls, many organizations define standardized reporting and control processes applicable to all programs."[5] Metrics identified and developed jointly between the senior management team and the program governance personnel are used as part of the monitoring and control for each program and represent the unique needs for that organization.

Those governance metrics and measures, some of which are identified in this chapter, will be normally categorized to cover:

- Planning: priority setting, funding allocation
- Execution: progress tracking, early warning indication
- Completion: assessment of deliverables, benefits
- Results Evaluation: closed loop correction, strategy assessment

Every firm's senior management, along with personnel responsible for administering their governance processes, is faced with the challenge of identifying the right metrics and achieving balance between effective control and monitoring with the overhead cost required to administer them. The investment in time needed to capture data and use meaningful metrics must be balanced against the value obtained. If, as stated earlier in the chapter, what gets measured gets improved, organizations must guard against measuring and monitoring the wrong or superfluous aspects of the program.

A SYSTEMS APPROACH TO METRICS

The research from John Hauser and Florian Zettelmeyer concluded that the use of metrics is key to program and project success.[6] Companies using consistent, balanced, and mutually aligned metrics outperform companies who use sporadic, schedule-oriented, and nonaligned metrics. The use of program metrics is an institutionalized practice of leading companies. They focus on regular, planned, and periodic measurements of program and business performance. So what does it mean to have a balanced and aligned set of metrics?

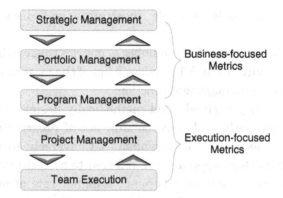

Figure 8.2 Using the IMS to balance and align metrics.

To accomplish this, one must take a systematic approach to design-ing a performance management system and choosing the metrics to use. In Chapter 3 we introduced the integrated management system (IMS), which describes the critical management functions used to conceive and deliver new capabilities to realize the mission, strategic goals, and busi-ness benefits of an enterprise. The IMS provides an effective framework for defining the business- and execution-focused metrics needed to mea-sure the performance of an organization's management functions in a balanced and aligned manner (Figure 8.2).

Metrics in support of the strategic management activities indicate per-formance toward achieving the strategic business goals desired. These metrics are highly business-focused and are used as input for business planning and setting of long-term strategy.

Portfolio management metrics are also highly business-focused and provide information on the effectiveness of the portfolio of programs and indicate if a business's capital and human resources are invested and allocated properly for the achievement of the business returns desired.

Metrics at the program level are a mixture of both business- and execution-focus, which indicates the role of program management as the *glue* that binds execution and strategy. In general, program management metrics provide an indication of performance against achievement of both the business benefits desired and the program objectives established during program definition.

Project management and team execution metrics are primarily execution-focused and provide information on performance toward the planning and execution of their deliverables. They are typically focused on measuring project-level triple constraints and discipline-specific outcomes.

There are a seemingly infinite number of metrics that can be employed. For a quick reference, Table 8.1 provides a list of the most common metric types for each IMS management function.

Companies that proactively and methodically design their program management approach, and corresponding set of program metrics, are what we refer to as best-in-class companies. Conversely, those that take an ad hoc, inconsistent approach are referred to as rest-in-class companies. Best-in-class companies experience a higher rate of program goal accomplishment and also produce a higher quality metric set—quality that is quantified by usefulness.

A detailed comparison between best-in-class and rest-in-class companies reveals that the best companies engineered and installed the use of metrics as part of a performance management system; the metrics were balanced and mutually aligned. Balanced metrics are those that cover all dimensions of a program, including schedule-oriented metrics, as well as those for financial, customer, process, and human resource utilization. Balanced metrics also include both leading and lagging indicators. Leading indicators are forward looking, such as a projected finish date based upon the rate of milestone completion. Lagging indicators, such as percentage of deliverables completed on time, are most valuable for a retrospective view of program performance. Both leading and lagging indicators are important and useful.

Mutually aligned means that metrics are compatible and use the same baseline information—for example, performance to planned schedule, probability of completing the program by a certain date, and the cumulative percentage of milestones that are accomplished are all based on the same baseline: the program timeline.

Literature claims that this behavior, using balanced and mutually aligned metrics, is not by chance, but rather is by design and is aimed toward enhancing success in terms of accomplishing the program goals.[7] Consequently, best-in-class companies translate program and business success factors into specific metrics the program team can act on and also create incentives for accomplishment of the metrics. For example, the executive team at Lucent Technologies studies the best time-to-market performance measures in the industry, which are turned into targets for their own teams to beat. This is a culture of continual improvement by making program targets highly aggressive.[8]

Table 8.1   Aligning metric types to IMS management functions.

IMS Management Function	Metric Types
Strategic Management	Revenue growth—from existing products
	Revenue growth—from new products
	Revenue growth—in existing markets
	Revenue growth—in new markets
	Product patents per year
	Reputation—Net Promoter Score®
	Employee engagement
	Most/Least effective product
	Most/Least effective market
	Most/Least costly customer
	Operational efficiency
Portfolio Management	Capacity—Number of programs and projects
	Average return on investment
	Pipeline throughput
	Complete versus canceled programs and projects
	Diversification of programs—by risk (portfolio risk index)
	Diversification of programs—by product type
	Diversification of programs—by customer targets
	Development turnover rate—Approval rating at each phase
	Ideas to market completion rate
Program Management	Staffing—Actual versus planned
	Funding—Actual versus planned
	Funding—Estimate to complete
	Timeliness/Schedule—Percent of deliverables met on time
	Complexity—Number of interdependencies
	Time-to-Market (timeliness target)
	Time-to-Volume (market share target)
	Time-to-Payback (financial return target)
	Projected future income from program
	Projected customer adoption (rate of new market share)
	Risk exposure—probability of realizing expected business benefit

(Continued)

Table 8.1   (*Continued*)

IMS Management Function	Metric Types
Project Management	Requirements traceability index (from plan through test)
	Performance against cost
	Performance against budget
	Performance against quality
	Performance against schedule
	Risk—probability of achieving deliverables on time, in budget

In summary, best-in-class organizations carefully build metric systems that align program execution with the organization's business strategy. In that effort, the emphasis is on using metrics to measure performance on a consistent basis from inception to completion of the program. They insist on measuring multiple facets, balancing metrics to obtain a holistic picture of program health, and selecting metrics that are aligned and compatible.

MEASURING BUSINESS BENEFIT

If the primary reason for program management is the delivery of business benefits, then program management metrics need to be concentrated toward measuring business results. This is the main reason why we make the distinction between *business* success factors and *critical* success factors.

In practice, we have observed that there is a strong tendency to use execution-focused measures when program managers first begin identifying their business factors. Examples are performance against schedule and performance against budget. This mistake is certainly understandable because many program managers are former project managers where the measurement focus is on execution, not necessarily on business performance. Additionally, current literature is dominated by reference to execution-focused measures and metrics.

Keep in mind, however, that a program manager serves as the business manager's proxy, and a business manager is most interested in measuring business results. When defining program-level metrics, we always recommend that a program manager focus on business metrics first, *and then* identify the critical operational metrics needed.

For clarification, consider the following example. Liberty Tax is a tax preparation and planning company operating in North America. The company funded a program named Sapphire to develop a new web-based tax preparation service for small business clients. The market introduction or go-live date was a critical *business* success factor. If the service was late to market, Liberty would miss all or part of the tax preparation cycle for the year. This would have a negative impact on the business benefits anticipated such as new customer acquisition and increased revenue (top line growth). In this example, as noted, the market introduction date is a *business success factor* for the Sapphire program. Additionally, the project managers chose to use *performance against schedule* as a key performance metric to provide cross-program concentration on timeline performance at the operational level. This demonstrates an alignment of metrics at the program and project level.

Program-level metrics are driven by the business benefits desired, so they will be unique for each program. Table 8.2 provides a catalog of program metrics that we commonly observe organizations utilizing in their practices.

It is easy to see how one can get overwhelmed by metrics. The metrics introduced thus far are simply a listing of those most commonly used. There are hundreds of metrics available for use. We are often asked, "How many metrics do I need for my program?" The answer is: *It depends*. A study by *BusinessWeek* and *Boston Group* shows that the sweet spot is somewhere between 8 and 12 metrics (see "How Many Metrics Do You Need?").[9] Our experience in managing both small and large programs supports this assertion. At the program level, using more than 12 metrics can cause more information overload than value.

How Many Metrics Do You Need?

The answer to this question of course depends on the nature of the program. The strategy, culture, and stakeholders' values of the company, as well as how big, complex, and technically and commercially risky a program is, all play a major role in deciding how many metrics you need. Additionally, accelerated time to money objectives for the program will require special attention to how the program is managed and what metrics will best assist management in keeping it on track. As a general rule, one needs a sufficient number of metrics to maintain good control

Table 8.2 Metric options relative to business benefits.

Business Benefit	Metric	Measurement
Financial	Return on Investment	(Gain from investment minus cost of investment) divided by cost of investment
	Return on Capital	Net income divided by (interest bearing debt plus stockholders' equity)
	Return on Assets	Net income divided by total assets
	Return on Net Assets	Net income divided by net assets
	Return on Sales	Net profit after taxes divided by net sales
	Net Operating Revenues Ratio	Income before other items divided by adjusted net operating revenues
	Profitability Index	Present value of future cash flows divided by initial investment
	Inventory Turnover	Cost of goods sold divided by inventory
	Gross Margin	(Sales minus cost of goods sold) divided by sales
	Net Margin	Net income divided by sales
Market	Market Share	Total company sales divided by total industry sales
	Target Market Index	Relative market size divided by (1 plus relative market growth rate)
	Market Coverage Index	Number of countries the organization sells in weighted by size of revenue) divided by target global market revenue
	Customer Fallout	Number of failed sales conversions divided by total possible sales
	Customer Acquisition Cost	All sales and marketing costs for a specific period of time divided by the number of new customers in the same period of time
	Time to Payback	Customer acquisition cost divided by how much the customers paid on average per month
	Incremental Sales	Revenue generated by marketing initiatives divided by baseline sales
	Net Marketing Contribution	(Sales revenue times percent gross margin) divided by marketing and sales expense

(Continued)

Table 8.2   (*Continued*)

Business Benefit	Metric	Measurement
Customer Satisfaction	On-Time Delivery	Number of orders delivered on time divided by the total number of orders received
	Customer Satisfaction	Number of customers noting complete satisfaction divided by total number of customers
	Net Promoter Score	Number of customers likely to recommend our company to a friend or colleague divided by number of customers asked
Quality	Product Defect Rate Assessment	Total number of products completed in a specific period of time divided by number of rejected products due to defects over the same period of time
	Defects Per Unit	Total number of defects found on all units divided by the total number of units
	Material Quality	Total orders with material quality within agreed tolerances divided by total number of orders
	Perfect Order Index	Percent on time multiplied by percent complete multiplied by percent damage free multiplied by accurate documentation
Business Productivity	Time-to-market	Average time from approval to launch for each product
	R&D Success	Number of new products launched over a specific time divided by the number of projects developed in that same period of time
	Cost Performance Index	Earned value divided by actual cost
	Schedule Performance Index	Earned value divided by planned value
	Percent of New Technologies	Number of new technologies produced divided by the number of planned technologies for the period

of a program, and as few as possible to still stay nimble. How many is that for an average program?

Some examples may provide guidance in answering the question of how many. In one successful high-tech company, program teams have eight standard metrics agreed upon with senior management. In another company, programs use between 10 and 14 metrics. In a leading semiconductor company, very complex manufacturing building programs evaluate about 20 metrics each month. Our personal experience is that about ten metrics are sufficient, if an effective metrics system is constructed.

CHOOSING METRICS THAT MATTER

Each company will have a distinct set of metrics depending on its business strategy and goals. Therefore, a set of metrics will vary by the type of industry the company is part of and by its business strategy (for example, differentiation, cost leadership, quality, etc.). The job of customizing a metrics set falls to the senior leaders of an organization and is often negotiated with program managers.

As an example, Frog (a leading product strategy and design firm) uses four common success metrics for their programs:[10]

1. Ability to bring a design to market
2. Client satisfaction
3. Profitability
4. How the program team feels about the work completed

While the job is ultimately that of senior management, when a business determines its own distinct set of metrics, it should do so in collaboration with company stakeholders and be in sync with the organization's governance process, procedures, and other requirements. Each stakeholder will emphasize different metrics according to their needs.

- The board of directors, financial community, and CEO will primarily be interested in the strategic metrics (see "I Have Only Three Minutes a Month!").
- Business managers will show a strong interest in metrics that assess the strategic health of the business, the alignment of programs with the business strategy, and the balance of the program portfolio relative to the business needs.

- Program management directors and governance board members will be concerned with metrics assessing the achievement of business benefits and operational health of a program and its constituent projects.
- Members of the IPT core team and the project teams will be most keenly interested in metrics about project management and team execution.

The variation in stakeholder needs has to be considered and comprehended when choosing program metrics. At times, trade-offs have to be made in the choice of metrics to prevent an overburden of information capture and dissemination.

I Have Only Three Minutes a Month!

Al Petroff, CEO of DirectConnect, a producer of interface cable, was very expressive when describing the kind of performance measurement report he prefers: "My biggest problem is time. I have 40-plus programs going at any one time. Some in early stage exploration, some in development, and some in operations. So, I need a set of metrics that will show the most important things about my programs, and I must be able to read it in three minutes."

Petroff assigned the task of designing such a metric set to his PMO director, Diedra Simmons. "The challenge was the three-minute time constraint," explained Simmons. "I knew that the metrics I chose had to be specific, understandable, useful, and timely."

Simmons designed a one-page metrics report covering program management-related measurement from strategic management, portfolio management, and program management perspectives. It featured five metrics: 1) performance planned business benefits, 2) program portfolio distribution, 3) projected future income from the program, 4) external customer satisfaction, and 5) percent of program milestones accomplished.

According to Petroff, "Diedra did a great job of identifying the metrics that were most important for running my business. In most cases the report even meets my three-minute rule. The exception of course is when a program is in trouble and I need to gather additional information."

Using Strategy as a Compass

As noted, metrics must be aligned with the specific strategy and objectives of the firm to be effective. As a reminder, it is the business strategy that dictates the configuration and focus of a company's program management

Table 8.3 Strategy drives metric decision—differentiation strategy.

	Company A
Company Business	High-tech electronics
Business Strategy	Differentiation
Type of Programs	New product development
Metrics (First Priority)	Time-to-market
	Percentage of new technologies
	Profitability index
	Market share
Metrics (Second Priority)	Customer fall-out
	Manufacturing cost
	Development cost

function. This means that each element of program management needs to be aligned with the business strategy and reflected by the program management metrics used. To illustrate this alignment, we will look at three companies with three different business strategies as defined by Michael Porter's model, and demonstrate how the strategy drives the selection of the metrics.[11]

Company A in Table 8.3 uses a differentiation strategy to offer its customers something different from their competitors. In particular, Company A focuses on technology innovation and fast time-to-market to achieve differentiation and gain higher profits and market share. Secondarily, Company A strives to achieve high quality and does not allow runaway cost. To measure performance to their strategy, Company A selects metrics for measuring time-to-market, percentage of new technologies developed, profitability index, and market share. Additionally, they choose customer fallout and development cost metrics to measure their secondary strategy of providing high quality and cost containment.

Company B focuses on a low-cost strategy aimed at establishing a sustainable cost advantage over its rivals. It competes in the manufacturing of fasteners used in building construction and uses a low-cost strategy to underprice rivals and capture market share. This strategy will leave Company B with a small gross profit margin per product, but profit can be achieved through a large volume of sales and lean staffing. Metrics to support this business strategy should measure development and manufacturing costs, market share, program staffing level, gross profit margin, and sales volume. Secondary objectives are to bring the product to the market in an average industry standard time and with average quality.

Table 8.4   Strategy drives metric decision—low-cost strategy.

	Company B
Company Business	Commodity fastener manufacturing
Business Strategy	Cost leadership
Type of Programs	Development programs focused on cost reduction and module integration
Metrics (First Priority)	Development cost
	Manufacturing cost
	Market share
	Staffing level
	Gross profit margin
	Sales volume
Metrics (Second Priority)	Development time
	Defect rate assessment

This is complemented with the secondary priority metric of development time and defect assessment rate. Table 8.4 summarizes the alignment of strategy and metrics.

Company C aims to become a low-cost provider with high-quality features and focuses on a best-cost strategy by combining upscale features with low cost. A secondary strategy is to develop innovative information technologies within reasonable time frames. Primary metrics to measure this strategy should include defect rate assessment, customer fallout, development cost, and manufacturing cost. Secondary metrics could include percent of new technologies delivered and development time. Table 8.5 summarizes the alignment of strategy and metrics.

Table 8.5   Strategy drives metric decision—low-cost/high-feature strategy.

	Company C
Company Business	Provider of automotive information technology
Business Strategy	Best cost and innovation
Type of Programs	Hardware and application SW supporting automotive features
Metrics (First Priority)	Defect rate assessment
	Customer fall-out
	Development cost
	Manufacturing cost
Metrics (Second Priority)	Development time
	Percent of new technologies to company

These examples illustrate how business strategy drives the choice of metrics. We have shown three examples of companies with different strategies and the corresponding sets of metrics chosen. Each company in the world will have its own business strategy and set of metrics to measure the effectiveness and achievement of its strategy.

Metrics Quality

As noted at the beginning of this chapter, metrics can do as much harm as they can good if they are not chosen and used properly. We express the need for quality when using metrics in order to overcome the common errors of metrics management and performance management systems. Best practice companies design their metrics systems as minimalistic and simple as possible; yet balance few and simple with the need for decision-making information. Sometimes, few are trumped by many because of the program and business need for the metrics.

As a quick reference guide (and mnemonic tool) to help overcome limitations associated with metrics, follow these "SMART" guidelines from George T. Doran.[12]

Specific: Objectives and metrics need to be clear and unambiguous. All stakeholders should be able to (consistently) define the metric relative to what is expected, why it is important, and how it is measured.

Measurable: The objectives must be timely and measurable. Therefore, data must be available for use and consistent to track (and trend) over time.

Attainable: Objectives must be realistic. Goals and objectives should include some that are high probability of occurrence and some that are a stretch for achievement.

Relevant: Goals, objectives, and metrics must be aligned from strategy through execution. In other words, they must be relevant. Additionally, to be relevant, metrics must be regularly used, otherwise the time to collect and illustrate metrics is not worth the investment of time and energy.

Time-Bound: Metrics must be grounded and the most natural way to do so is by time. Time provides a sense of urgency, especially if corrective action is needed in between metric reports.

Many organizations over-compensate when using metrics. Organizations not "SMART" about using metrics often find they are committing one or more of the common errors noted at the beginning of the chapter. When the errors occur, frustrations cascade from projects to programs to the organization at large. Quality in the use of metrics relates to those that are the fewest needed, the simplest to manage, and are used regularly in discussions and for decision making. Additionally, quality in metrics includes those that have a clear owner and point of responsibility.

Metrics and Decision Making

There are hundreds of decisions to be made during the course of a program—some large and significant, some small and incremental, and some that fall between these two extremes. It is well documented that having access to the right data at the right time facilitates effective decision making.[13] Difficulty in finding information to support a decision, or even the perception of difficulty, can affect decision quality on a program. Decision makers often trade off decision quality versus time to find the information needed.

Having an effective metric system for a program serves to bring the data and information needed close to the people in charge of making decisions on a program. Having the *right* metrics is critically important, however.

Selecting the metrics to use to drive information-based decisions involves determining the data needed to measure and quantify a decision—or more accurately, the decision criteria. Further, decisions at the program level must support the business goals and program objectives. Therefore, a tried and true approach for selecting one's metrics remains valid for program management.[14]

1. Establish the goals and objectives.
2. Determine the necessary decisions and decision criteria.
3. Select metrics that will measure the criteria.

Utilizing this approach ensures that both program decisions and metrics support the goals and objectives of the program.

ENDNOTES

1. Lombardi, M., and Saba, J. Talent assessment strategies: A decision guide for organizational performance. Aberdeen Group, 2010. Retrieved September 1, 2013 from www.cpp.com/pdfs/Aberdeen.pdf.

2. Fenton, N. E., and M. Neil. "Software Metrics: Roadmap." *Proceedings from Conference on the Future of Software Engineering*, pp. 357–370, 2010.

3. Daniel, D. R. "Management Information Crisis." *Harvard Business Review* 39(5): 111–116, 1961.

4. Hauser, J. R., and G. M. Katz. "Metrics: You Are What You Measure!" *European Management Journal* 16(5): 516–528, 1998.

5. Project Management Institute. *The Standard for Program Management*, 3d ed. Newton Square, Penn., 2013.

6. Hauser, J. R., and F. Zettelmeyer. "Metrics to Evaluate R, D, & E." *Research Technology Management* 40(4): 32–38, 1997.

7. Ibid.

8. Meyer, C. "How the Right Measures Help Teams Excel." *Harvard Business Review* 72(3): 95–103, 1994.

9. McGregor, J. "The World's Most Innovative Companies." *BusinessWeek Online*, April 24, 2006.

10. Vander Meere, R., personal interview, November 27, 2013.

11. Porter, M. E. Competitive Strategy: *Techniques for Analyzing Industries and Competitors*. New York: Free Press Publishers, 1998.

12. Doran, G. T. "There's a S.M.A.R.T. Way to Write Management's Goals and Objectives." *Management Review* 70(11) (AMA FORUM), pp. 35–36, 1981.

13. Dennis, Alan R. "Information Exchange and Use in Group Decision Making." *MIS Quarterly*, December 2006.

14. Tillman, Frank A., and Deandra T. Cassone. *A Professional's Guide to Decision Science and Problem Solving*. Saddle River, N.J.: FT Press, 2012.

Chapter 9

Program Management Tools

Program management tools support the practices and various processes used to effectively manage a program. They are enabling devices for the primary players on a program—the program manager, the project managers, the executive leadership team, and the governance body.

The tools presented in this chapter do not represent a comprehensive suite of program management tools. Rather, they are a key subset of tools we have found to be most impactful and widely implemented within best practice organizations. These tools are both strategic and operational in nature and include:

- The Benefits Map
- Program-level Work Breakdown Structure
- The Program Map
- Complexity Assessment
- The P-I Matrix
- The Program Strike Zone
- Indicators and Dashboards

When used properly, program management tools are a useful aid for decision making. Each of the tools presented provides important information needed to make critical program-related decisions by the primary program management players identified above. Their function is to help establish and maintain alignment between business strategy and program execution, and to facilitate effective program execution in order to realize the business benefits intended.

THE BENEFITS MAP

Business benefits are the outcomes of a program that provide value to the organization in return for the investment made in the program. The benefits map documents the expected benefits that are to be realized from the investment. It specifically charts the path from the organization's business strategy to the distinct benefits that are to be derived from the output of a program and each of its constituent projects.[1] The information provided by the benefits map supports the cost/benefit analysis for a program and should become part of the program business case.

Tool Description

The benefits map can be depicted in visual form to provide a useful means to demonstrate alignment of project deliverables, to program objectives, to the business success factors for a program, to the expected business results as displayed in Figure 9.1.

As demonstrated in the figure, the benefits map provides traceability between project outcomes and deliverables to the benefits intended from the output of the program. This is critical information to first establish the overall vision and scope for a program, then communicate how each of the constituent projects contribute to the goals of a program, and finally trace the execution of the program and projects to final delivery of business benefits.

Like many program-based tools, this tool is used to assist in characterizing how specific program objectives are met. However, benefits maps can become complex and confusing due to the *one-to-many* relationships between project deliverables and outcomes to the program objectives. The critical component in building an effective benefits map is to ensure each project deliverable or outcome is mapped to a program objective, and every objective to the business success factors. For more complex programs, it is recommended that the mapping be performed and represented in tablature form to establish order, reduce confusion, and maintain greater value than effort in creating the map.

Utilizing the Tool

Along with the program work breakdown structure and the program architecture, the benefits map is a useful tool for establishing the overall

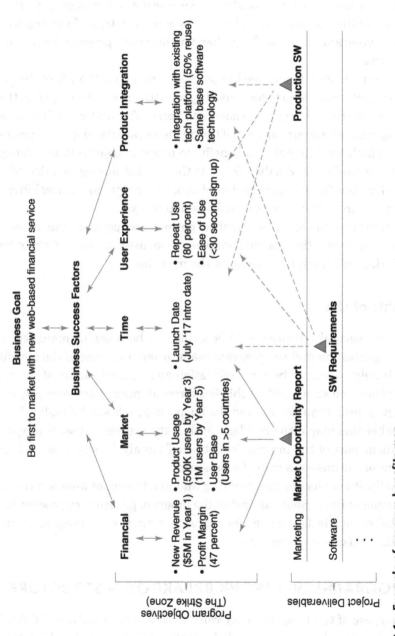

Figure 9.1 Example of program benefits map.

scope of a program and for demonstrating the alignment between project outcomes and deliverables to the objectives of a program. The benefits map can also be used to communicate to senior leaders, the overall program team, and other program stakeholders how the strategy of the organization and program are melded together, and how each program benefit will be realized.

The benefits map is intended to be used throughout the life of the program to analyze consequences caused by adjustments and changes as they occur to the original program vision and scope. The first use of the benefits map normally occurs as part of the business case development process where a high-level mapping of benefits to program objectives to strategic intent is established. Further detail is then added during detailed planning when the full comprehension of program scope and traceability of project outcomes to business benefits is necessary.

In organizations with established governance policies and a set of related processes, the benefits map is often used as one of their key monitoring and tracking devices for the program.

Benefits of Use

There are several advantages to be gained by both senior management of an organization and the program team through the use of the benefits map. It helps to create better clarification and understanding of the program vision and scope, and establishes direct alignment between program objectives, project outcomes, and the business benefits to be realized.

The benefits map also provides a systematic process to assess program benefits as part of the program's cost/benefit analysis, which is a critical element of the business case of a program.

Finally, it enables focused tracking and monitoring of progress toward realization of the benefits as part of the program governance process, and establishes an effective means for evaluating success of a program from a benefits realization perspective.

PROGRAM-LEVEL WORK BREAKDOWN STRUCTURE

The purpose of the Program-level Work Breakdown Structure (PWBS) is to translate the elements of the whole solution being created and delivered by the program into tangible work outcomes. It is a comprehensive characterization of program scope, including scope of the constituent projects.

It specifies *what* is to be done, not *how* or *when*. If something is not included in the PWBS, it is not needed to achieve the business results desired and therefore should not be included as part of the program effort.

Tool Description

The PWBS is one of three key elements of program management that determines the total scope of the program. The other two elements are the program architecture and the benefits map.

Normally, a PWBS consists of three or four levels of detail as illustrated in Figure 9.2. Level one describes the whole solution, which the program is chartered to create. Level two identifies the physical and nonphysical elements documented in the whole solution. Most often this level identifies the projects and project enablers that make up the program. The information for level two is derived directly from the program architecture. Levels three and four display the deliverables or outcomes that result from the work of the projects and enabling representatives involved on the program. Project deliverables constitute tangible work output (final evaluation report), whereas a project outcome identifies the completion of work (evaluation complete). The deliverables and outcomes defined by the PWBS form the basis of the deliverables and outcomes used in the benefits map (Figure 9.1).

Just like the project-level WBS, the PWBS does not include the element of time or sequencing of activities. It simply focuses on the identification of the deliverables and outcomes generated by the projects and functional enablers.

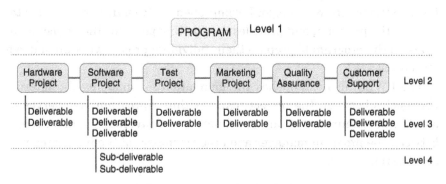

Figure 9.2　Example of a PWBS.

Utilizing the Tool

The PWBS is used to decompose the program's output into more manageable components that will serve as the starting point for developing the work breakdown structures for each of the projects within the program. It is normally first established as part of the detailed scope definition, but *after* the program architecture has been established.

As demonstrated in Figure 9.2, different branches of the PWBS may have varying degrees of detail. That is, greater depth of deliverables may be required for some projects as compared to others. Complexity of the program and each of its projects is the determining factor on the appropriate level of work decomposition.

A common error that occurs when developing a PWBS is mixing both deliverables and tasks. A PWBS should only include deliverables or work outcomes, not tasks. The PWBS does not eliminate the need for a detailed work breakdown structure for each of the constituent projects on the program. In practice, both are required.

Most program teams find that the PWBS is a useful tool to be used between the program manager and the project managers for coordination, communication, problem resolution, and ensuring that all activities are identified and synchronized as planned. It is particularly helpful during conversation regarding interfaces between projects and other internal and external functions involved with program mapping.

Benefits of Use

The PWBS is a critical tool for properly defining the scope of a program and for managing program complexity. The tool offers two key advantages to the program manager. It comprehensively characterizes the total scope of the program, and provides the starting point for the project managers on the program to prepare their individual project work breakdown structures.

The PWBS also establishes the foundation for mapping the interdependencies between the project teams within a program, and serves as an aid in communication and coordination of deliverables and outcomes between the program manager and other members of the integrated program team.

THE PROGRAM MAP

The program manager is responsible for managing the interfaces between the projects within a program. The program map is an essential program management tool that helps the program manager (and project teams) visualize the interfaces over the life of a program.

Its purpose is to identify critical cross-project interdependencies with respect to deliverables and time. The crux of the program map is to show the big picture view of the cross-project deliverables and help the program manager and program team fully understand the dependencies that exist on the program.

Tool Description

The structure of the program map, as well as the deliverables for each project, is derived directly from the PWBS described earlier. This forms a tight connection between the program architecture, PWBS, and the program map.

Each deliverable or outcome identified in the PWBS is displayed on the program map in the correct sequence and point in time. The use of arrows between deliverables depicts the interdependencies between the project teams and their respective deliverables. A simple example of a program map is illustrated in Figure 9.3.

In this example, "Pwr Contl SW" is a deliverable generated by the software development project team that is "delivered" to the enclosure project team for the development of the enclosure. In turn, the "Enclosure Design Files" deliverable is generated by the enclosure project team and delivered to the manufacturing project team. This mapping of deliverables shows the critical interdependency between the software, enclosure, and manufacturing project teams of the program.

The criticality of the interdependencies between the project teams on a program is easily demonstrated with this tool. In the above example, failure to complete any of the deliverables on the part of the software or enclosure project teams will have a devastating effect on the work of the manufacturing team. In fact, it will have a devastating effect on the entire program, thus the need to visualize the interdependency of deliverables on a program.

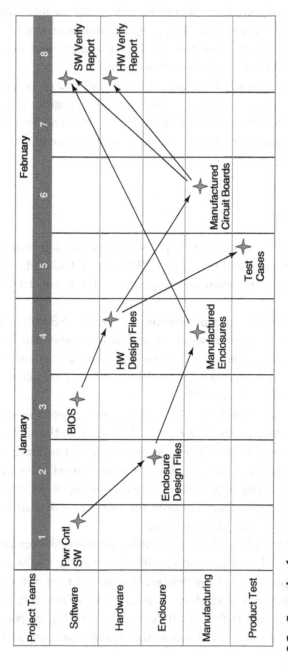

Figure 9.3 Example of program map.

Utilizing the Tool

For any program—small, medium, or very large—the program map is highly recommended to fully understand the deliverables and interdependencies between the project teams on the program. The best time to create the program map is during the beginning of the planning cycle, but only after the program scope and requirements have been defined. A work session (or workshop) approach for creating the program map yields an information-rich exchange, cross-project discussion, face-to-face interactions, and reciprocal iterations that have no equal when it comes to developing a quality program map.

If the facilitated approach is used to develop the map, one work day is a typical time duration, although very large programs can take multiple days. If a cross-project work session is not used, it is recommended that a full work week be allotted to complete a program map for small to medium-sized programs, and multiple work weeks for large programs.

It is important to note that a key input to the program mapping exercise is the program decision points that are identified as part of the program governance system. Once the dates for the program decision points are identified, they become the guiding schedule constraints for the mapping process. A word of caution to the program manager: even though a timeline is developed as an outcome of the program mapping exercise, this timeline should *not* be taken as the program schedule. A more detailed analysis of risks, resources, and budget is required before creating the program schedule. The program map simply serves as a sequence of events over time, not the actual schedule.

Benefits of Use

As stated previously, the program map helps the program manager and program team fully understand the dependencies that exist on the program. Use of the tool also offers the following benefits:

- It provides clarity and visibility to the program manager and team for the critical cross-project interdependencies that must be managed.
- It provides an integrated display of the planned deliverables and work outcomes of all project teams within the program.
- It provides a high level view of program execution.
- It serves as the basis for developing a program timeline.

Due to the necessary investment of time and program resources to develop a program map, the program manager must decide when sufficient information is available and reasonably stable to create the map. This will minimize the number of iterations of the map.

PROGRAM COMPLEXITY ASSESSMENT

Complexity is an attribute of all programs. Designs have become more complex as features and integrated capabilities increase; the process to develop and manufacture the solutions is requiring more partners, suppliers, and others throughout the value chain; the ability to integrate multiple technologies with end user wants requires not only accuracy regarding requirements delivery, but also requires speed and agility to change; and the current global, highly distributed business environment requires work to occur in multiple sites across multiple time zones. Therefore, the ability to characterize and profile the degree of complexity associated with a program has become essential for both executive leaders and their program managers.

The information gained from the use of a complexity assessment tool helps to balance the portfolio of programs from a complexity perspective. Further, it aids in the determination of the skill set and experience level required of the program team; it guides the implementation of key program processes such as change management, risk management, and contingency reserve determination; and it helps the program manager adapt his or her management style relative to the level of complexity of the program.

Tool Description

The structure of the program complexity assessment features several parts. First, the tool includes various dimensions as defined by a business. Each dimension of complexity is assessed on an anchor scale (second part), and when the complexity scores of each dimension are connected, a line called the complexity profile (third part) is obtained. The complexity profile is a graphical representation of a program's multifaceted complexity. An example of a program complexity assessment is illustrated in Figure 9.4.

Every industry has discrete characteristics, every business within an industry is unique, and every program within a business has unique

Complexity Dimension	Low Complexity		1	2	3	4		High Complexity
Business Climate	Stable				X			Uncertain
Market Novelty	Derivative				X			Breakthrough
Financial Risk	Low			X				High
Objectives	Clear			X				Vague
Tech. Requirements	Clear			X				Vague
Organization	Hierarchical				X			Matrix
Technology	Low-tech			X				Super Hi-Tech
Speed to Market	Regular			X				Blitz
Geography	Local					X		Global
Team Members	Experienced	X						Inexperienced

Figure 9.4   Example of program complexity assessment.

attributes. This means that a firm has to customize the complexity assessment tool for its use.[2] Program managers often start this work by determining the complexity dimensions that are specific to the organization. For example, technical complexity may be directly related to the technical aspect of the product, service, or other capability under development, or from the knowledge and capability of the existing resources of the firm. Structural complexity also has a number of subfactors that involve the organizational elements of a firm. Business complexity involves the business environment in which the firm operates.

Examples of variations in dimensions with respect to technical, structural, and business complexity are shown in Table 9.1.

Utilizing the Tool

Typically, the program complexity assessment tool is prepared during the program definition and planning work cycles—that is, very early in the program. However, this tool should be used dynamically and updated periodically in high-velocity environments where the program scope and business climate may frequently change. It is advantageous if the

Table 9.1   Example technical, structural, and business complexity dimensions.

Technical Complexity	
Low Complexity	**High Complexity**
Feature upgrade to an existing product	New product architecture and platform design
Development of a single module of a system	Development of a full system
Use of existing and developed technologies	Use of new and undeveloped technologies

Structural Complexity	
Low Complexity	**High Complexity**
Team is co-located	Team is a virtual team
Mature processes and practices	Ad hoc processes and practices
High performing team	Low level of team cohesion
Single-site development	Multi-site development
Single-geography development	Multi-geography development
Single-cultural team	Multi-cultural team
Single-company development	Multi-company development

Business Complexity	
Low Complexity	**High Complexity**
Selling into traditional and mature markets	Selling into new and emerging markets
Receptive customers and/or stakeholders	Unreceptive customers and/or stakeholders
Flexible time-to-money requirements	Aggressive time-to-money requirements
Existing end-user usage models	New end-user usage models

senior management team of an organization that manages the portfolio of programs uses this tool to approve the overall level of complexity for each program as it may directly influence their decisions on the skill level and experience of personnel required for each of their programs.

As stated earlier, each organization should create a customized version of the assessment tool that is specific to the complexity dimensions they are dealing with. Once the complexity dimensions are identified, each dimension is then assessed based upon the scale established. For example, in Figure 9.4, speed to market is assessed as a level-2 complexity (fast and competitive). Once all complexity dimensions are assessed, connect the obtained scores for each dimension to produce the complexity profile that helps to visually depict the overall program complexity. The profile in Figure 9.4, for instance, indicates that the program is of medium complexity, with all dimensions at levels 2 and 3, except team members who are experienced (the least complex) and a globally distributed team (the most complex).

The senior management team and program manager can quickly get a feel for the level of complexity of each program within the organization. However, care should be taken to prevent the inclusion of too many complexity dimensions. In this case, the simpler the structure of the tool, the more effective its use will be.

Benefits of Use

The program complexity assessment tool's value is multifold. First, knowing the program complexity helps balance the portfolio of programs with an appropriate mix of low-, medium-, and high-complexity programs (investments). Further, the complexity assessment aids in the planning process, indicating how to adapt one's management style to the level of complexity of the program.

The program complexity assessment tool also helps the senior management team determine the level of skill and experience needed, thus aiding in the selection of the program manager and the key leadership positions on the team to successfully define and execute the program. Additionally, the tool may influence how much contingency buffer to build into the program budget and schedule—the more complex the program, the bigger the buffer. Finally, the tool can help identify the categories of risk and the level of robustness you will need in your risk management plan.

THE P-I MATRIX

The P-I (Probability-Impact) matrix is a tool used to identify program risks, assess their probability and potential impact, and to provide a representation of risk severity to facilitate effective risk-based planning and decision making.[3] The P-I matrix is used by the program manager as part of the risk management process and practices discussed in Chapter 7 for managing program risk throughout the life of the program.

Tool Description

The P-I matrix is normally presented in a matrix format with the vertical axis reflecting the *probability* of a risk event occurring and the horizontal axis displaying the projected *impact* if the risk event occurs. The higher an identified program risk is on the probability axis and the further to the right on the horizontal axis, the greater the severity of risk to the program.

The values for *probability* range from very unlikely (1) to nearly certain to occur (5), while the values for *impact* range from very low (1) to very high (5).[4] The position of a program risk on the matrix determines its estimated *severity* value through the sum of the combined values assigned for its probability and impact. Figure 9.5 illustrates an example P-I matrix.

The matrix is usually divided into red, yellow, and green zones, which represent high-, medium-, and low-severity risks respectively, based on the organization's thresholds for risk severity. The position of a risk in the matrix determines its ranking or severity. The higher the value of a square, the higher its rank and the more severe its potential impact to the program. To be effective, the matrix must be realistic (as to the severity of the risk), timely, and brought to life by the joint effort of the program and project managers who own the P-I matrix.

Utilizing the Tool

One of the challenges for a program manager is to identify the risk events that have both the highest impact on the program and those that are most

PROBABILITY (P)	Risk Score = P + 2 x I					
NC = 5	7	9	11	13	15	High Severity
HL = 4	6	8	10	12	14	Medium Severity
L = 3	5	7	9	11	13	Low Severity
LL = 2	4	6	8	10	12	
VU = 1	3	5	7	9	11	
	VL = 1	L = 2	M = 3	H = 4	VH = 5	
	IMPACT (I)					

Figure 9.5 Example of P-I Matrix with low, medium, and high severity.

Table 9.2   Example five-level scale of risk impact on schedule.

Scale	1 Very low	2 Low	3 Medium	4 High	5 Very high
Risk Impact on Schedule	Slight Schedule Delay	Overall Program delay 5 %	Overall program delay 5–14%	Overall program delay 15–25%	Overall program delay >25%

likely to occur. Therefore, the impact, probability, and severity of each program-level risk need to be analyzed. In this assessment, we tend to use a numeric probability scale. For example, on a five-level scale, the ratings are 1 = very unlikely, 2 = low likelihood, 3 = likely, 4 = highly likely, and 5 = nearly certain. Consequently, one will qualitatively assess each risk's probability of occurrence. When this is completed, the next step is to assess the potential impact of each risk, again on a discrete scale. For example, a scale rating such as 1 = very low, 2 = low, 3 = medium, 4 = high, and 5 = very high may be used. To illustrate, let's assume a risk that will be assessed as having a program schedule slip impact. The scale can define the levels of impact, as shown in Table 9.2.

After all program-level risks are assessed in this manner, a formula can be used to combine the probability and impact of each risk to establish a measure of severity. Although nonlinear formulas can be employed, linear formulas such as $severity = [(probability + N) \times impact]$ are easier to apply. For example, N can be equal to two, meaning that impact is twice as important as probability in establishing risk severity. In this case, the assessed probability and impact for each risk would be entered in the formula, $severity = [(probability + 2) \times impact]$, and the obtained value would be entered into the P-I matrix. This is the formula utilized for the risk severity calculations in Figure 9.5.

Some larger programs commonly focus on the top ten highest-ranked risks. In contrast, some smaller programs decide to manage the top three risks, arguing the lack of resources prohibit taking on a larger number of risks. Both approaches are dangerous. So, what is a reasonable way out? The answer is in the P-I matrix—respond to the highest-ranked risks in the matrix, down to an agreed upon severity level.[5] For example, focus on handling risks down to risk score of 11 (see Figure 9.5) and treat other risks as noncritical. With this approach, one neither squanders resources nor disregards significant risks. It should be noted that noncritical does not mean not important. Rather, it means that scarce program resources are not immediately needed to address the risk event but may be needed in the future.

It is important to keep in mind that program planning and execution tools and the information and analysis derived are not static. The environment during the time span of a program is constantly changing as is the risk associated with a program. Therefore risk must be constantly evaluated.

Benefits of Use

The P-I matrix helps sift through the myriad of uncertainties to pinpoint and highlight the program areas of highest risk—both before work has begun and throughout the life of a program.[6] This offers an opportunity to focus program resources and to identify effective ways of reducing risk events in a proactive manner, rather than being confronted by them if they turn into problems later in the program. In addition, the matrix generates information for more reasonable contingency planning and effective program decision-making (Chapter 7).

It is impossible to predict all possible risk events. Therefore, best practice organizations include the use of schedule and budget contingency based upon a risk analysis. The amount of contingency can be derived from risk impact information contained in the P-I matrix.

THE PROGRAM STRIKE ZONE

The program strike zone is used to identify the critical business success factors for a program, to help a program manager and his or her executive management team track progress toward achievement of the key business results anticipated, and to set the boundaries within which a program manager and team can operate without direct management involvement. The program strike zone name is analogous to a baseball pitcher seeking the "sweet spot" for throwing a successful strike. In the business world, the program manager is attempting to hit the sweet spot of business success through successful delivery of a program's output capability.

The tool fosters a "no surprises to management" philosophy and behavior by increasing the flow of relevant information between the program team and executive management. This assists in providing an efficient means for elevating critical issues and barriers outside the span of control of the program manager for decision making and resolution by senior management. Additionally, the program strike zone provides a mechanism for stopping a program if it is no longer aligned to the objectives of the company.

Tool Description

The program strike zone is an effective tool for ensuring the program is planned with the correct set of success criteria, and that the program stays within the success criteria boundaries throughout the life of a program.

As shown in Figure 9.6, elements of the program strike zone include the critical business success criteria for the program, target and control (threshold) values, and a high-level status indicator. Some organizations also include an "actual" element that provides indication of where a program is operating with respect to the target and threshold limits.

The business success factors will be unique to every organization, and are derived directly from strategic management and portfolio management processes (Chapter 3). The strike zone is most effective when the business success factors identified are kept to a critical few (usually 5–6), as this focuses the program and senior management's attention on the highest priority contributors to the success of the program.

Program Strike Zone			
Critical Business Success Factors	**Strike Zone**		**Status**
Market Share: Increase market share in product segment	**Target**	**Threshold**	
• Order growth within 6 months of introduction	10%	5%	Green
• Market share increase one year after introduction	5%	0%	Green
Time-to-benefits Target:			
• Program Initiation Approval	1/3/2017	1/15/2017	Green
• Business Case Approval	6/1/2017	6/30/2017	Yellow
• Integrated Plan Approval	4/15/2018	4/30/2018	Green
• Release to Customers	5/30/2018	6/15/2018	Green
Financials			
• Program Budget	100% of plan	105% of plan	Green
• Product Cost	$8500	$8900	Red
• Profitability Index	2.0	1.8	Green

Figure 9.6 Example of program strike zone.

The threshold limits are negotiated values between the program manager and the executive sponsor. They establish the empowerment boundaries within which a program team is free to operate without senior management intervention.[7]

The status indicators give a high-level indication of how the program is operating within the target and threshold values. A green status indicator signifies progress is as planned, yellow status indicates a "heads-up" to management of a potential problem, and red requires management intervention in order to proceed.

Utilizing the Tool

The program strike zone provides many uses to program managers, the executive sponsor of a program, and to the program governance body. Program managers use it to formalize the critical business success factors for the program, to negotiate and establish the team's empowerment boundaries with executive management, to communicate overall program progress and success, and to facilitate various trade-off decisions throughout the program cycle.

Executive managers use the program strike zone to ensure that a new program's definition supports the intended business objectives, and to establish control limits in order to ensure that the program team's capabilities are in balance with the complexity of the program. When used properly, it provides senior managers a forward-looking view of program alignment to the business objectives. When problems are encountered, the tool's structure is intended to provide an early warning of trending problems, followed by a clear identification of "show-stopper" conditions based on the level of achievement of the critical business success factors. If a program is halted, senior executives can either reset the critical business success factor target or thresholds, modify the scope of the program to bring it within the current targets, or in the worst case, cancel the program to prevent further investment of resources.

Executive managers and the program governance body set the boundary conditions (targets and thresholds) of the program strike zone within which the program manager can operate, thereby empowering the program manager to make decisions and manage the program without direct senior management involvement. For example, if a program manager and team are well seasoned and experienced, senior executives may choose to "open up" the difference between the target and threshold values, which effectively gives the program manager a higher degree of empowerment. Conversely, if a relatively inexperienced program manager and team are involved, or a program is highly complex and risky, senior executives may choose to tighten the difference between target and threshold values. This in effect establishes stricter governance control over a program.

As long as the program progresses within the strike zone of each critical business success factor, the program is considered on target and the program manager remains fully empowered to manage the program through its life cycle. However, if the program does not progress within the strike zone of each critical business success factor, the program is not considered on target and the executive managers must directly intervene.

Benefits of Use

In practice, the value of the program strike zone is achieved through the direct communication and interaction between the program manager and the executive sponsor of the program in setting the vision and key success parameters. These parameters are then recorded in the strike zone and become the management and tracking focus for keeping the program aligned to its business and operational objectives.

Use of the tool fosters a "no surprises to senior management" behavior by increasing the flow of relevant information between the program team and executive management. This results in an efficient means of elevating critical issues and barriers to success for rapid decision-making and resolution.

When used appropriately, it enables empowerment of the program manager and team by establishing the boundaries for authority, responsibility, and accountability. Too often we hear program managers tell us that they have all the responsibility for driving program success, but lack the authority. The program strike zone is the best tool we are familiar with to balance both sides of this equation.

Bill Shaley, a senior program manager for a leading telecom company, described the program culture within his company this way: "Managing a program is like having a rocket strapped to your back with roller skates on your feet, there's no mechanism for stopping when you're in trouble." Sound familiar? The program strike zone is such a mechanism. The strike zone is designed to stop a program, either temporarily or permanently, if the threshold limits are breached, at which point the program is evaluated for termination or reset for continuation.

INDICATORS AND DASHBOARDS

The program manager spends a great deal of time reporting the status of their program to various stakeholders. Regardless of whether the status report is formal or informal in nature, the message should remain constant. The means by which this is accomplished is by consistent, comprehensive, and accurate information collected from the project team leaders and the remainder of the IPT members.

The following tools are used to effectively communicate business and operational status and progress at the various levels of a program.

The Project Indicator

The project indicator is used to establish cross-project reporting of project-level status during IPT core team communications. Program managers must be specific on what elements of the projects they want reported, the metrics to be used, and the level of detail to include. By working with the project managers in developing a consistent format for the indicators and a constant set of metrics to report, the program manager will receive a comprehensive yet concise report on project status in a standardized format. This will enable the program manager to efficiently evaluate operational progress on each of the projects and to consolidate the information for reporting program-level progress to senior management and the program governance board.

Tool Description

A project indicator is a brief, one- to two-page report that shows pertinent project status and trend information. An example is illustrated in Figure 9.7 for a software development project that is part of a larger program.

The project indicator format and data to be used is agreed upon between the project and program manager. There are certain elements in a project indicator that will be unique to each project, such as software bug tracking in this example, and other elements that will be common for all projects, such as status, schedule, and risks.

Utilizing the Tool

The IPT core team should engage in a review of status for each of the projects on the program on a regular basis. For many programs, this is

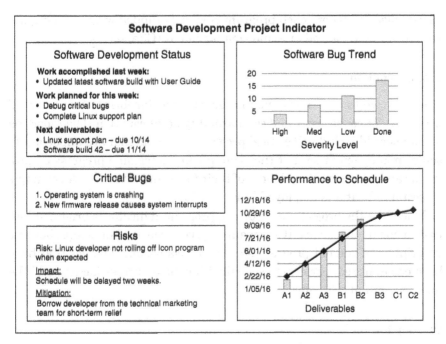

Figure 9.7   Example of project indicator.

normally performed weekly, but in some cases it might even be performed on a daily basis, such as in an agile software environment.

Reporting the status and updates through the project indicator facilitates the flow of information within the program and keeps the team focused on the top priorities for achieving the program objectives. It also provides important information on which projects within the program require focused attention on the part of the program manager.

The Program Indicator

In most organizations, the program manager is required to provide status to senior management on a regular basis. The program indicator is used by the program manager to summarize the overall status of a program based upon the input and discussion with the IPT core team. The tool gives the program manager a high-level view of the total program and helps the program manager to determine if the program remains successfully on track or if there are potential barriers and issues that must be addressed.

Depending upon how formal reporting is accomplished within the organization, the program indicator is useful in an organizational program review meeting where each program currently funded within the organization will be reviewed by senior management (Chapter 7).

Tool Description

It is effective to have a common program indicator format for use on all programs for consistency and comparability of information. The reporting format should include all critical program elements that are important to senior management so that they can quickly determine progress on programs and which ones need more of their focus and attention. An example program indicator is displayed in Figure 9.8.

Much like the project indicator, the program indicator is brief and limited to one or two pages. It is meant to give a concise but comprehensive description of current program status, key issues and changes that have been encountered, performance against business success factors, and the management of critical risk events.

Utilizing the Tool

The program indicator is normally used as a formal report that is distributed to senior management and program governance members as part of program review meetings. It can be used effectively with the program strike zone and program dashboard to communicate both business and operational program status. The objective for the program manager is to communicate both strategic and operational information regarding the program to the executive level, and to use the forum to gain executive assistance when needed.

Benefits of Use

The program indicator provides the project managers on a program a consistent format and data model for determining the current status of their program, understanding the most critical challenges their program team is facing, and a gross-level indication of how their program is trending against plan.

The program indicator also serves as a key communication vehicle between the program manager and the senior executives that highlights key cross-project issues that need to be elevated to senior leadership for resolution.

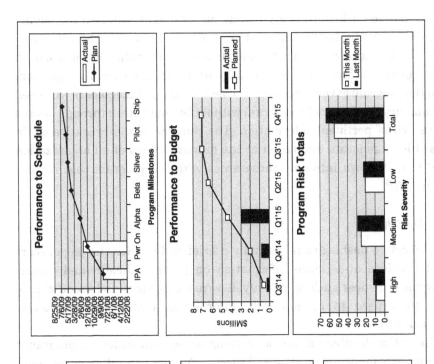

Program Indicator

Program Overview
- Circuit board power on complete
- Currently 4 weeks behind schedule
- Currently $1.2 million under budget
- Gap in engineering resources
- Memory support issue resolved

Program Status

Work accomplished last month:
- Alpha customers identified and committed
- Circuit board power on complete
- SW build 42 delivered
- Validation team staffed and fully tasked
- Marketing plans completed
- Enclosure CAD files delivered to vendor

Current Issues
- Validation platform stability
- Four weeks behind schedule on evaluation
- Critical part shortage for next circuit board builds
- Currently a five week gap in engineering resources.

Performance to Schedule

Program Milestones

Legend: Actual, Plan

Performance to Budget

Legend: Actual, Planned

Program Risk Totals

Risk Severity

Legend: This Month, Last Month

Figure 9.8   Example of program indicator.

233

The Executive Dashboard

An executive dashboard is a business management tool used to visually communicate the health and status of a program from a business perspective. Creation of the program dashboard is a function of the program management office, while the primary users are the executive sponsors and the program governance board. These are busy stakeholders who need a summary level view of a program without having to troll through details in search of pertinent information. The purpose of the executive dashboard is to generate discussion at the executive level of an organization and to drill down into only those areas within a program that require attention.[8]

Tool Description

An executive dashboard is normally a brief and concise one-page report that provides at-a-glance information on program health and status. Graphical devices such as green/yellow/red indicators and trend charts are often used to communicate information clearly and succinctly (Figure 9.9).

Executive dashboards should be designed to communicate a program's core metrics for health, status, and success. A program's intended business outcomes and an indication of a program's progress toward achieving the critical business success factors should always be included. If used, a program strike zone will provided detailed status of business success factor health, including the green, yellow, and red indications. This information can then be summarized and included in the executive dashboard.

An indication of the overall program risk level and progress toward management of the critical risk events is valuable information for determining program health. Detailed information on risk severity can be found in the P-I matrix and other tools that can be summarized and included in the executive dashboard.

Consideration should also be given to the inclusion of information on changes in program scope, quality management, program dependencies, resources, and customer or organizational readiness to receive the program output.

Utilizing the Tool

The executive dashboard is used by the executive sponsor or the program governance board to build awareness of status and drive critical decisions

Program Name	Status	Financials (YTD GM)		Financials (Budget)		Introduction Date		Market (Orders)		Customer (Sat.)	
Program 1	Green	Tgt. 20.0%	Est. 21.4%	Tgt. $1.0M	Act. $7.54k	Tgt. 4/22/16	Est. 4/15/16	Tgt. $102.7M	Act. $50.3M	Tgt. 7 (out of 10)	Act. 6.2
Program 2	Yellow	Tgt. 15.5%	Est. 15.4%	Tgt. $2.5M	Act. $2.38k	Tgt. 9/12/17	Est. 9/20/17	Tgt. $80.0M	Act. $68.9M	Tgt. 7 (out of 10)	Act. 5.1
Program 3	Green	Tgt. 31.0%	Est. 29.7%	Tgt. $742k	Act. $121k	Tgt. 7/29/19	Est. 7/29/19	Tgt. $98.1M	Act. $0.32M	Tgt. 7 (out of 10)	Act. N/A
Program 4	RED	Tgt. 24.0%	Est. 25.9%	Tgt. $3.0M	Act. $3.62M	Tgt. 6/05/15	Est. 6/30/15	Tgt. $222.3M	Act. $220.0M	Tgt. 7 (out of 10)	Act. 5.9

Green: On track to meet program goals
Yellow: Warning, program goal(s) in jeopardy
Red: Stop, program goal negatively impacted, management intervention required

Figure 9.9 Example of executive dashboard.

related to the program. In general, the dashboard is designed to provide answers to the following questions:

- Is the program properly aligned to business strategy?
- Is the program on target to achieving the business success criteria?
- Are there deviations from the original plan and success metrics?
- Are there critical risks and issues affecting the program?
- Have there been significant changes in the program or program environment?
- Does the program team need anything from senior management?

Based upon the information contained in the dashboard, the stakeholders may decide to intervene to provide advice and coaching, provide staffing or other types of support, or change program requirements as needed. If program objectives are compromised or potentially compromised, more serious actions may need to be taken including termination of a program.

Executive dashboards are updated at regular intervals depending upon an organization's reporting cycle. Normally we see the dashboards updated on a monthly basis, but depending on the program, updates can also occur biweekly or quarterly.

Benefits of Use

The value of the executive dashboard is in the direct communication and interaction between the program manager and the executive sponsor of the program in understanding the health and status of a program from a business perspective. The dashboard format provides the opportunity for executive stakeholders to quickly understand the current situation relative to a program and identify areas that need immediate attention and improvement. It also increases the visibility of risks and issues faced by the program team, thereby increasing the opportunity for early intervention and proactive management of the program's success factors.

USING THE RIGHT TOOL FOR THE JOB

Like all tools, program management tools are not meant to be a panacea that will automatically make someone a better program manager. Becoming a better program manager is accomplished through experience and continuous building of skills and competencies. Rather, tools are designed to help a program manager become more effective and efficient in performing the key program management practices. Using the right tool for the right job, however, is a necessary factor in gaining the desired assistance from a tool.

Each tool described in this chapter has a primary purpose and usage. The benefits map is used to establish alignment between project outcomes and deliverables to the business benefits of a program, while the complexity assessment is used to characterize and profile the complexity associated with a program.

The PWBS is used to develop a comprehensive characterization of program scope by decomposing the program's output into more manageable components that will serve as the starting point for developing the work breakdown structures for each of the projects within the program. Along with the PWBS, the program map is used to visualize the big picture view of the project deliverables and outcomes over time, and help the program manager manage the cross-project interdependencies that exist on the program.

The program strike zone is used to identify the critical business success factors for a program and the P-I matrix is used to identify the various risks to hitting the criteria. It is used to assess probability and potential impact of program-level risk and provides a representation of risk severity to facilitate effective program decision making.

Project and program indicators are used to aggregate information on the current state of each project on a program as well as the program holistically. Program dashboards are used to provide a high-level summary of current state of all programs within a portfolio and help to focus attention on issues and needs as they surface on a program.

In addition to the tools described in this chapter, we have included other tools throughout this book which also have specific usages. The alignment matrix (Chapter 3) is used to establish the alignment of program objectives to the strategic business goals driving the need for the program. The program business case (Chapter 6) is used to describe and detail the business opportunity and obtain the funding and resources needed to execute a program. Stakeholder maps and stakeholder assessment tools (Chapter 7) are used to develop a concise and targeted stakeholder strategy.

As stated previously, the tools introduced in this book are not meant to be a comprehensive inventory of tools that are available for use by a program manager. Additional tools and tool templates can be found on the Program Management Academy website: http://wiley.program management-academy.com.

Program Management in Practice

Agile Program Management

"Should we use program management or agile methods to plan and lead our strategic initiatives? Which is better and why? By the way, my name is Tim Cummings, director of product development. Welcome to the firm."

These were the questions directed to Robert Redding in his first meeting on his first day of work. Redding was hired to help lead strategic initiatives for a cloud-computing company that provides customer relationship management services.

Redding was an experienced leader in the industry with nearly twenty years of practice-based knowledge in both program management and agile software development. He was neither baited nor fazed by the questions. In fact, his hiring manager, Kate Williamson, explained to him during the interviewing process that the firm was struggling to differentiate the two disciplines. She warned, "There is an ongoing debate here with two rather entrenched camps, one promoting program management, and the other promoting agile development. Our director of product development, Tim Cummings, is a strong proponent of agile." Part of Redding's job was to help the company align efforts and bridge the divide between camps.

Redding was introduced to others at the meeting and first provided some background about himself and his role at the company. Before answering the questions posed to him, he began by asking a question in order to gain additional understanding, "What is the problem you are really trying to solve?"

Cummings was quick to respond, "We are putting product features out to market at a furious rate to satisfy our user's requests, but for the second year now we have failed to hit our business goals."

Elisa Saghari, VP of platform operations, joined the conversation. "The way I see it, that is exactly the problem. We are executing to satisfy customer direction, and not paying enough attention to executing to business goals." She went on to explain that their agile software development processes were resulting in products that were developed from the ground up. The platform, for which she is responsible, has become a collection of project outcomes that, when aggregated, do not represent the true intent of the product vision and business goals.

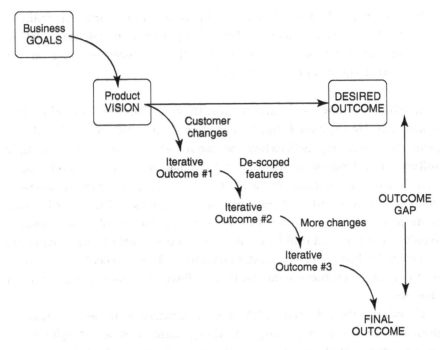

Figure 9.10   Business goals and project output misalignment.

Saghari then sketched an illustration for Redding (Figure 9.10).

She explained further, "Our products are how we achieve our business goals, but a problem is occurring during project execution. Because our agile project teams are chartered to respond to customer requests and are self-directed teams, we are experiencing a gap between our *desired* business outcomes and our *actual* business outcomes." She then followed with, "We need a way to keep our execution aligned to our business goals."

Redding recognized that what Saghari had described was a recurring problem for companies that are highly software reliant. He had learned that the two most common causes for the misalignment between agile project outcomes and business goals are as follows:

1. Not all requests for changes from customers are in alignment with a firm's business goals. In fact, some changes may be in direct conflict with the business goals.

2. If agile sprints and releases become time and resource constrained, any descoping decisions needed to keep a sprint or release on schedule may result in elimination of critical features that deliver the product vision and business goals.

Redding was now prepared to answer the questions that Cummings had asked him. He explained that both agile software development and program management practices have proven to deliver business results. Agile software development adds value to the business by quickly and continuously delivering product features to one's customers. Program management provides continued focus on business results and the coordination of multiple project outcomes toward the achievement of those business results. He explained that by combining the two practices, firms can gain a powerful solution delivery system that helps keep project teams focused on business goal achievement, yet remain flexible to customer and market changes.

Cummings then chimed in. "I've been approaching this as an 'either-or' decision. We either adopt program management or we adopt agile development. You are telling us we need to find a way to use both."

Redding was direct. "Yes. It's Agile Program Management."

As the conversation continued, Redding further explained: "There exists a serious misunderstanding of Agile Program Management and how it is successfully implemented and practiced within organizations. It is quite common for Agile Program Management to be confused with agile project management as well as with traditional portfolio management practices."

Redding then explained, "To fully succeed from a business perspective, there must be alignment from business goals to daily agile project outcomes. Alignment begins with a set of business goals that the firm is trying to achieve over a specified period of time. The product vision must support attainment of the business goals along with the roadmap of product releases." Seeing expressions of agreement on the faces of the people he had just met, Redding continued. "From an execution standpoint, agile releases, sprint outcomes, and daily outcomes must all support the product vision. Program management is what keeps this all in alignment."

It was from this meeting, the first one on his first day with the company, that Redding recognized the full extent of his role to inform and educate the firm on how to implement agile program management.

Creating Awareness

A dynamic business environment exists in most every industry. In fact, the business climate in which most companies operate today is changing so rapidly that within a short period of time, both business goals and the project execution plans that support those goals can change dramatically. In some businesses, if six months pass without a significant change to plans, there is a high probability that something was missed during planning or there is not enough attention and understanding of the market. Therefore, there is a high probability that something will be missed in project execution and the project outcomes.

This is especially true in any business where software development is a key component of solution development. The challenge that this environment creates is the need to be able to respond quickly to changing climate and customer needs, while at the same time being able to maintain a level of structure to keep work output aligned to the company's business goals.

As Redding worked with Cummings and his product teams, he consistently emphasized that agile programs use most of the same constructs that traditional programs use such as focusing on business goals, managing business risk, and relying on project interdependency maps (program maps) to name a few. The manner in which the program team operates is consistent with agile principles, with the cornerstones of agility applied at the program level. The difference is that Agile Program Management delivers business value early and often through the life of the program while maintaining continuous focus on business goal achievement and integration of multi-project outcomes.

Redding focused his conversations on emphasizing five key elements of Agile Program Management: alignment, collaboration, cadence, transparency, and integration.

Alignment: Business goals provide the basis for establishing alignment among the project teams associated with a program. When multiple agile project teams are needed to create the various elements of a solution, a common vision has to be in place to continuously align the work and outcomes of each of the teams.

Collaboration: On any program, a high level of collaboration is required due to the cross-team interdependencies that exist. Agile programs are no different. The agile project teams need to collaborate with one another to validate assumptions, determine and track cross-team commitments, and fully understand the program business requirements. Collaborative methods include release planning, sprint reviews, and retrospectives. Release planning involves the whole team in estimating the product backlog, deriving an understanding of the business requirements, and determining the final priorities. Sprint reviews create an opportunity for stakeholders to interact directly with the program team to validate and touch what was developed. Retrospectives are used to improve the development process and are a key tool for agile teams to increase throughput.

Cadence: The cadence of work and outcome delivery is set at the program level and aligned to the cadence of the organization as a whole. Each team may have different iteration lengths, but they need to be synchronized at regular synchronization points so that iteration planning and reviews can be held at the same time on the program. These sync points are also used to set expectations with stakeholders to keep them informed and actively engaged with the program.

Transparency: Agile teams do not fear bad news because they understand that the sooner an issue is realized, the cheaper it will be to fix. Transparency means that the whole program can see and touch what is in development early and often. This provides stakeholders with enough information early in the cycle to validate the business case and make any course corrections when necessary.

Integration: With multiple agile project teams working concurrently to develop and deliver a solution, integration of work output from each of the project teams is a critical component of Agile Program Management. As the agile project teams pull from the list of business objectives and customer requirements to establish team level product backlogs and release plans, it is the combination of all the team level release plans that constitutes the program's overall release plan. It is the program management function that ensures that the outcomes contained in the program releases integrate effectively to create the intended product or solution.

Cummings and his product teams began to examine their practices to determine how they might modify work elements to accommodate both program management and agile development. At one point Cummings asked Redding, "If we were to do one thing differently in an attempt to move toward an Agile Program Management model, what should we focus on first?"

"Focus on the business. Align your agile outcomes to business goals," was Redding's reply.

Agile Program Management in Practice

Two years later, Cummings tells a story of success. "Agile Program Management has emerged from the need to improve the management of our large strategic initiatives. An agile framework centered on scrum was selected to unify development cadence and release cycles. However, we realized that scrum by itself is not enough to align multiple scrum teams on common business goals."

The company still deploys three software release cycles a year. Within each release, all scrum teams synchronize on a monthly sprint boundary where it is expected that each team demonstrate the potentially releasable software increment developed in that iteration.

The monthly boundaries are critical to the process and are used as sync points for executive stakeholders and product management to examine the work in progress and realign priorities based on what is being shown. Product backlogs are fluid and change often but never impact what a team is working on in any given iteration.

Each scrum team is responsible for creating a bottoms-up iteration plan, selecting onto their backlog the features they will deliver and how they will develop them. Each team estimates the backlog at a high level to get a relative sense of the scope of effort for the whole release then proceeds to break down the backlog into an iteration plan for development. The program management function provides oversight of the release management process.

Program managers lead the integrated release planning process across scrum teams, keep track of cross team and cross functional dependencies, and manage program level risks. They help accomplish the attainment of the anticipated business goals from a leadership position in which multiple agile project teams operate to produce the outcomes necessary to deliver the business goals. Figure 9.11 illustrates the Agile Program Management model graphically.

Agile Program Management begins with the identification of the strategic business goals by the senior leadership team of the firm. Strategies, in the form of products and services, are then developed and assigned to program managers to develop and deliver. Delivery is accomplished through

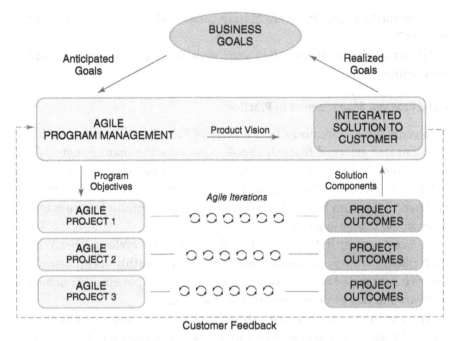

Figure 9.11   Agile program management.

the formation of multiple agile projects, each chartered to develop and deliver its appropriate portion of the product or service.

Each agile team has a good understanding of the project goals and features and establishes their team release plans, affirms dependency timeframes, and adjusts overall priorities based on more detailed understanding of the backlog. Development activities occur incrementally and continue until the release goals are met. The end of each iteration marks the point where agile project teams and the overall program team look back and retrospectively determine what changes need to be made to the plans, processes, or even to the product itself based on what was just developed in terms of a potentially shippable increment. These iteration reviews represent excellent opportunities for stakeholders, customers, and agile program teams to examine the working product, review planning assumptions, and recalibrate the product backlog to changing customer needs and business goals.

Each agile team in the program stays aligned through regular synchronization points facilitated by the program manager. In addition, there are program-level core team meetings to monitor and manage program risk, review overall progress, and establish transparency for executive stakeholders. The program core team is responsible for keeping agile project teams focused on the development goal and helps to protect them from outside distractions.

During one of his recent staff meetings, Tim Cummings pointed out that while once thought to be on opposite ends of the flexibility and structure continuum, agile software development and program management practices are now combined to create a flexible framework to develop and deploy solutions that are aligned with and help achieve the business goals of his company.

ENDNOTES

1. Sasghera, Paul. *Fundamentals of Effective Program Management: A Process Approach Based on the Global Standard.* Plantation, Fla.: J. Ross Publishing, 2008.
2. Edmonds B. "What Is Complexity?" In F. Heylighen and D. Aerts, *The Evolution of Complexity.* Dordrecht: Kluwer, 2006.
3. Concurrent Engineering User Group (CEUG). *Benchmarking Study.* Detroit: September 2005.
4. Vose, David. *Risk Analysis: A Quantitative Guide,* 2d ed. Hoboken, N.J.: John Wiley & Sons, 2000.
5. Milosevic, Dragan Z., and P. Patanakul. "Standardization May Help Development Projects." *International Journal of Project Management,* Vol. 23, No. 2, 2005.
6. Meredith, Jack R., and Samuel J. Mantel, Jr. *Project Management: A Managerial Approach,* 6th ed. Hoboken, N.J.: John Wiley & Sons, 2006.
7. Martinelli, Russ, and Jim Waddell. "The Program Strike Zone: Beyond the Bounding Box." *Project Management World Today,* March–April 2010.
8. Treasury Board of Canada. Guide to Executive Project Dashboards, website: www.tbs_sct.gc.ca/itp-pti/pog-spg.asp.

Part IV

The Program Manager

To this point in the book we have explained how program management is an extension of business strategy; how a program is defined, planned, executed, and terminated; and what common practices, metrics, and tools are available for use by the program manager.

This part of the book contains two chapters that focus on the program manager, specifically, the program manager's primary roles and responsibilities, and the competencies and skills that a program manager needs in order to effectively manage programs on a consistent basis.

In Chapter 10, "Program Manager Roles and Responsibilities," we focus on the three overriding roles of the program manager in program-oriented organizations: 1) manager of the program business, 2) integrator of multi-project work, and 3) leader of a cross-discipline team. All three roles are necessary to lead a team of multi-discipline specialists in creating and delivering a holistic solution, which becomes the means to achieving the firm's business goals.

To successfully respond to these three roles, the program manager needs to work toward proficiency in a broad set of competencies. In Chapter 11, "Program Manager Competencies," we present the program management competency model. The model contains the necessary knowledge, skills, and abilities to plan and execute programs proficiently. The competency model covers customer and market, business and financial, leadership, and process and project management proficiencies. Additionally, in this chapter we describe a number of organizational enablers that, when in place, support the program manager's growth and continued success.

Chapter 10

Program Manager Roles and Responsibilities

We now turn the focus of attention to the program manager. As we stated in Chapter 1, the role of the program manager can vary greatly from company to company based upon how an organization implements its program management function. As Figure 10.1 illustrates, program managers in project-oriented organizations tend to fill an operational support role that focuses on coordination and facilitation of activities, while program managers in program-oriented organizations take on an expanded role that also includes team leadership and business management. Additionally, since program managers within program-oriented organizations are typically responsible for delivering a whole solution (Chapter 4), they must also take on a large integration role.

In this chapter we focus on these three overriding roles of the program manager in program-oriented organizations: manager of the program business, integrator of multi-project work, and leader of a cross-discipline team. However, there is one important point we want to make before moving on. It is critical for the senior leaders of an enterprise to realize where they are operating with respect to the program management continuum and properly set the roles and responsibilities of their program managers accordingly.

THE BUSINESS MANAGER

Companies *survive* by generating enough income to pay for their operating expenses. Companies *grow* by investing a portion of their income

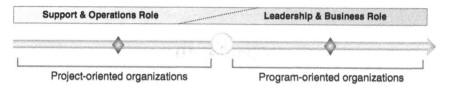

Figure 10.1   Variation of program manager roles.

in vehicles that generate more revenue and decrease operating expenses. Companies *sustain their growth* by developing and executing strategies to obtain long-term increases in profitability. Therefore, the income a company uses to fund its long-term growth is the investment that is used to generate an intended return. For program-oriented companies, programs and their respective outputs are the strategic means for achieving their sustained growth and business viability.

This program investment model is analogous to a personal investment financial model. As an example, let's look at the investment situation for a private investor named Shannon. Shannon's salary is the income she uses to pay her living expenses and fund her growth in net worth. Without investing a portion of that income in higher-growth investment vehicles, her net worth grows very slowly. If Shannon wants to accelerate the growth of her net worth, she looks for investment vehicles that will generate additional income over and above her salary. If she develops a portfolio of mutual funds, for example, each fund becomes a vehicle for generating additional income. Collectively the portfolio of mutual funds Shannon owns provides continual positive ROI, if managed properly.

A single mutual fund within the private investor's portfolio is analogous to a single program that a business manager invests in to generate a positive return-on-investment for his or her business. Both a mutual fund and a program are vehicles for generating a higher rate of return. To extend this analogy, the private investor and the business manager both define short- and long-term goals they want to achieve and implement investment strategies for attainment of the goals. They both may turn their investment over to someone else to manage—the private investor employs an experienced mutual fund manager, and the business manager employs an experienced program manager. The mutual fund manager manages a collection of stocks that make up the mutual fund, and the program manager manages a collection of projects that make up the program.

As explained in Chapter 2, in many organizations the business manager cannot personally scale to manage the business on all of the programs

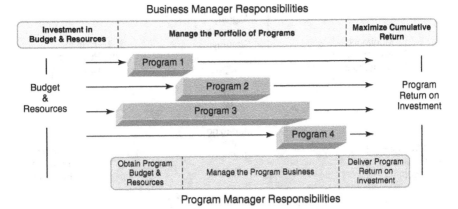

Figure 10.2　Shared business responsibilities.

within his or her portfolio, and therefore must allocate responsibility for managing the business aspects of programs to the program managers. The program manager in effect serves as the business manager proxy. Figure 10.2 illustrates this sharing of business responsibility.

As the business manager proxy for a program, the program manager assumes his or her first role, managing the *business* on the program. This is what a senior vice president in charge of a business unit for a major aerospace and defense company said about the role of the program manager in his organization:

> Program managers have one of the hardest jobs in our industry because they have to guide all business aspects of running their programs by walking a fine line between delivering profits to the company and delivering value to the customer.

Managing the Business

There are several elements the program manager is responsible for in order to manage the business of a program, as depicted in Figure 10.3.

Each of these seven elements of managing the program business is critical for a program manager who is operating in the business manager quadrant of the program management continuum (Figure 10.1). However, for program managers who are operating in other quadrants of the continuum, not all of the seven elements in Figure 10.3 are always relevant. Relevancy is determined by the expectations and definition of the program manager's role by the senior leaders of an organization.

Figure 10.3   Elements of managing the business.

Aligning to Business Strategy

The program manager normally serves as the primary business strategist on a program by continually focusing on the strategic business goals driving the need for the program and for its output.

We often see a common pitfall where strong alignment to the strategies of a business is not established prior to program business case development. Instead, the alignment is attempted somewhere between business case approval and capability release, creating a higher probability of misalignment between the program and strategic objectives. It is never a pleasant experience for the members of a business unit when they realize that the strategic return on their investment will not be realized due to a misalignment between strategy and execution.

During program execution the program manager continually monitors the business environment to ensure the program remains strategically viable. This is most important for programs with long cycle times, which have a greater risk of strategic shifts due to environmental changes. At any point in which misalignment between the program objectives and the strategic goals develops, the program manager has the responsibility for bringing it to the attention of the senior management team for discussion and resolution.

Managing the Business Case

The program business case establishes the business feasibility and viability of a program and is the primary tool for the program manager to manage the business aspects of a program (Chapter 6). Managing to the business case becomes a critical piece of the program vision and forms the basis of the business contract between the program manager and the senior management team of the company.

The role of the program manager is to drive the creation of the business case based upon the knowledge, information, and data available at the time. Since there are many unknowns to comprehend when a business case is created, establishing and documenting the set of assumptions that the business case is built upon is crucial. For the business case to remain accurate and viable, the program manager needs to continually manage and update the business case as additional information and data become available.

Part of the business case involves developing a robust set of business success factors and metrics that are used to determine the viability of the program. Collectively, the business success factors define business success for the program. They serve as the business objectives that the program team works to achieve, as well as the primary measure of checks and balances to guide decisions and to evaluate environmental changes. The program strike zone (Chapter 9) is a useful tool for managing a program to the business case.

At any point in the program cycle in which the business case no longer looks viable, the program manager must communicate that to the senior sponsors of the program. This, at times, can be difficult due to emotional and organizational culture factors, but as the business manager proxy on a program, the program manager must manage the return on investment. This usually means managing the program to success, but it can also mean cutting the losses as early as possible and recommending to terminate the program if the investment is no longer viable.

Managing the Program Finances

Once a program is approved or awarded funding, the program manager becomes responsible for managing the investment funds. The funds may come from various sources such as lending institutions, venture capital companies, shareholders, customers, or internal funds. This involves managing the program cash flow and managing the cost of goods sold. It is recommended that a program manager have a financial analyst on the IPT to help manage the financial aspects of a program, as managing program finances can be a time-consuming activity. Great benefits can be gained if the program manager forges a strong working relationship with his or her financial analyst (see "Knowing the Numbers").

Knowing the Numbers

Like some program managers, Michael, a product development program manager with Lockheed Martin, viewed financial analysts as a thorn in his side. But a conversation during a routine meeting with Sarah, the financial analyst on his core team, was the beginning of a million-dollar lesson that taught him the value of working closely with his finance team.

At one point during a discussion about cutting the cost of prototype systems that were sent to lead customers, Michael inadvertently asked Sarah where the revenue generated from the sale of the prototypes goes.

Because she had access to and understood the company's accounting system, Sarah was able to track the revenue. According to Michael, "during a follow-up meeting Sarah explained that the money goes into the general sales and marketing department, which isn't tied to our department." Michael then asked Sarah if it were possible to route the revenue back to his department, and, better yet, back to his program. "I wanted to see if the money could be brought back to the business unit to use for developing more products, or brought back to my program to alleviate some of the financial constraints we were working under," Michael said.

Sarah was able to work within the company's financial system to have the money routed back to the program once it was received from the customers. After the money had been collected and brought back into the program, Michael and Sarah presented their quarterly financial report to their management team. "For the current quarter, the program showed zero cost expenditure, and we were in the middle of the execution phase where maximum expenditure was occurring," stated Michael. Sarah explained, "We collected enough revenue from the sale of the prototype products to cover our development expenses for the quarter."

Obviously, the senior management team was pleasantly surprised, and they required every program thereafter to follow the same financial process. As a result of this and further cooperation between the program manager and financial analyst, the program finished a million dollars under the planned budget.

"It was a learning experience for me," Michael added. "There is positive value in working closely with a financial analyst if you think beyond the constraint piece of it. They're just doing their job—not giving you an open checkbook—and that causes contention between you and the analyst at times. But they are definitely an asset."

Managing the program cash flow involves more than just monitoring the total available program budget. It involves understanding and monitoring program expenditures and income on a periodic basis. All companies have to make money to stay in business, and the ability to maintain a positive cash flow is a critical component of an organization's success. This is especially true for smaller companies, start-ups, and any company that is cash constrained.[1]

Once a program is funded, management of cash flow for the program becomes the responsibility of the program manager. It is risky to monitor the program budget in total. If, for example, the budget was depleted three-quarters of the way through a program, the team would suddenly be in trouble with respect to program finances, and most likely work would come to a halt (at least temporarily, if not permanently). For this reason, the program manager along with the program financial analyst should create a cash flow plan for the budget and then manage program finances to the plan. When a periodic cash flow plan is established (Figure 10.4), deviations from the plan are much easier to react to and manage.

Understanding the billing and payment terms of the company, its suppliers, and its customers should be understood and reflected in the cash flow plan. There is a time delay for both expenditures and income generated by a program due to billing cycles. For example, the organization sponsoring a program may have payment terms of 90 days after receipt of goods from its suppliers. This means that expenditures for the program

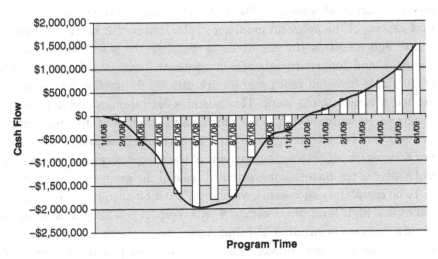

Figure 10.4   Example of program cash flow.

are delayed by three months. Likewise, revenue generated by the program will also be subject to the payment terms of a company's customers.

Knowing when to ask for assistance from the senior management team of the business is an important part of cash flow. Short-term and long-term program cash flow assistance can only be provided by senior management, and the program manager must be willing to work with the executive team to secure the finances needed to continue the program.

Managing the cost of goods sold is also a critical aspect of managing the program finances. The cost of goods sold includes the bill of materials and the cost associated with manufacturing, assembling, or otherwise physically creating a solution. The cost of goods sold is deducted from the revenue generated for each unit sold or used in the case of a service. Minimizing this cost enables a company to retain a larger percentage of the sales revenue or cost savings generated once the solution is delivered.

The program manager and his or her program team are responsible for driving the future cost of goods sold to a minimum. Many cost versus performance trade-off decisions are made during a program to generate an optimum cost of goods that achieves the performance requirements of the program. The program manager's role is to set aggressive, yet realistic, goals for the program team. Then he or she must measure progress toward achievement of the cost goals using program metrics (Chapter 8).

Managing Program Resources

Aligning a program to the strategic goals of a company, developing a viable business case, and managing the program finances are critically important aspects of the program manager's role. However, if he or she cannot secure and maintain the resources to complete the work, the strategic, business, and financial goals of a program will not be achieved. Much of the funds invested in a program are utilized to pay for the resources needed to complete the work. This includes both human and nonhuman resources. Resource cost is by far the largest consumer of the investment, in many cases constituting more than 75 percent of the program budget. For this reason, effective management of the program resources is a critical aspect of the business management role of the program manager.

To be completely successful, a program must be adequately resourced. There is a high level of uncertainty with respect to resource utilization when a program is initiated. For that reason, resource management needs to be viewed as a continuous process.[2] A common pitfall that we have observed is that a lot of effort goes into generating a program resource plan and gaining commitment from resource managers to fulfill the resource demand of the program. However, once the plan is approved, focus shifts to other aspects of the program, and resource constraints begin to mount. The program manager must manage resource utilization throughout the life of a program.

Resource management involves the program manager understanding the skills, experience levels, and number of resources needed for a program. The program manager is first responsible for ensuring the core team is fully staffed. This means each project has a qualified project manager in place and represented on the core team. Further, each critical function supporting the program should have a qualified representative. Secondly, the program manager is responsible for working with each project manager to be sure that his or her team is adequately staffed. When resource constraints are identified on a project, the program manager assists the project manager in resolving the constraint as effectively as possible and, if necessary, adjusting the project plan and program plan to take the constraint into account. Resolution usually entails adjustments to program costs, resources, time, performance, or quality goals. At any point in a program in which resource gaps adversely affect the program success factors, the risk must be communicated to the senior management team for assistance in resolution.

Managing Business Risk

The financial world teaches us that all investment involves some level of risk. Developing new capabilities and driving organizational change transitions are no different. Various levels of risk taking are necessary if a company is intent on being a leader, or even a strong contender in its industry. Market leaders understand that they must run directly toward, instead of away from, risk to put distance between themselves and their competitors. This means that programs can have a high level of business risk associated with them. Failure to identify and understand the risks, or uncertainties, can lead to substantial loss for the enterprise. The role of the program manager is to protect the company's investment in the program through effective management of the business risks involved.

Risk is inherent in every program—it comes with the territory. Good risk management will not make the risk associated with a program go away; it merely helps to ensure that the program team will not be blindsided with a problem to which they cannot respond.[3] The program manager must be aware that the failure of a single project within a program can cause the entire program to fail. This requires that the program manager empower the project managers to manage the project-level risks and focus his or her efforts on identifying and managing the program-level risks (Chapter 7). The role of the program manager is to be an advocate for effective risk management practices on the program and ensure that it becomes an institutionalized practice on the program. The program manager needs to keep the entire program team engaged in risk management practices, because risk management becomes useless if only a small number of program team members effectively practice it.

Managing Intellectual Property

Intellectual Property (IP) may have significant strategic and commercial value to a company. Much of the IP generated by companies engaged in product, service, or infrastructure development is done so within the context of development programs. For this reason, a crucial element of managing the business on a program involves the proper management of the IP created. Depending upon the type of IP created, this may include making sure that the proper patents, trademarks, and copyrights are obtained (see "Intellectual Property of Another Kind").

IP is the know-how that is produced by our creative minds. In particular, it is the codified content, data, and knowledge of doing something in a better way. It is also the knowledge to figure out that this better way may be very profitable if it can be commercialized. If a company patents its technology, for example, it would prevent others from using it without paying for its use.

Multiple players such as technologists, managers, functional managers, directors, legal experts, and officers all take part in managing IP. However, program managers must develop a sense for the following questions:

- *Is there a competitor who could use the IP developed on my program to compete against us?*

  This means that the program manager should know the market and customers and have a good understanding of the competitive landscape.
- *If so, what does that mean to my program's business?*

  Again, good knowledge of market, customers, and competitors, along with the program business case, will help the program manager understand the answer to that question.

- *What can be done to protect the IP?*

    There are multiple ways to protect IP. The trick is to determine the best approach for protecting the IP based upon its type.

The program manager should use this knowledge to ensure that all players involved play the game in a way that benefits the program because, ultimately, the program manager owns the program success.

Intellectual Property of Another Kind

Lois Lee, director of the PMO for Prdex, Inc., a high-tech manufacturer said, "One of the most important pieces of Prdex's IP is our program management process, although it is not legally protected as IP. I would suspect that we have made several hundreds of millions of dollars in profits through the use of our processes. Of course, I should have told you our history to understand my statement. We introduced our program management process 28 years ago.

"We have had five major revisions to our program management process over the 28 years on the basis of our collective learning. Further, all involved in programs (executives, engineering managers, program managers, core team members, and extended team members) received extensive training on the program management process.

"I guess that we have conducted 500-plus programs over the 28 years with the processes. Program managers come and go, but our program management process stays. It is a product of collective learning and know-how, and has enabled our company to introduce new products to market more rapidly and at higher financial returns than our competitors. This is why we consider it IP."

Monitoring the Market

Besides a strong internal focus on the business of the program, the program management role also requires an external focus on the market and continual monitoring of market conditions. The market consists of the customers who buy a solution, the end users who will use it, and the competitors who are trying to sell their solutions to the same customers. There are many examples of program managers who have focused their attention solely on the effective use of their resources to achieve the goals of the program—all the while being unaware of changes in the external market that have left their solution ineffective and unattractive to the customers and end users.

The best program managers continually monitor the state of the external environment to check if the business objectives and strategies that the program is based upon are still achievable. In particular, to keep a program feasible from the business and strategic perspective, the program manager should be fully aware of the current state of the market in which the business operates and also be aware of emerging trends. To determine the current state and trends of the market, the program manager needs to be engaged with the customer and bring the customer information back to the program team. It is crucial that program managers understand the wants and needs of their customers and end users to obtain business success for the program. For an example of a creative approach to garnering end-user information, see "From the Design Room to the Showroom."

From the Design Room to the Showroom

Shane Plaxton, a Hewlett-Packard (HP) research and development manager in Vancouver, Washington, knows first-hand the value of interfacing with end users. An Oregonian newspaper article explained HP's approach to putting its printer engineers and designers in direct contact with end users.[4] During a recent holiday season, HP sent more than 1,600 employees into electronics and office-supply stores to speak directly to end users, demonstrate how their products work, and offer their expertise.

Plaxton said the shopper feedback was a "little intimidating" because he is "not used to being out in front of customers." But he added that it was refreshing to step back from the details of printer design and see the big picture. "I spend all day looking at designs and product specs and details," he said. "This is a way of going out and seeing what end users really want to get."

The experience inspired him to take a fresh look at how HP can simplify its printer designs to make it easier for end users to choose the best model for them and to understand how it works.

Marketing experts say companies such as HP realize their product designers are too far removed from the end user and are working quickly to close the gap. Focus groups and customer surveys, the easiest and most popular marketing tools, are taking a backseat to more innovative approaches. Companies have discovered that observing end users and their decision processes in real-world situations gives a better picture of what buyers want.

HP workers each write a brief report on their store visits, highlighting end users' reactions to specific products. Jim Mury, the Vancouver manager in charge of HP's employee demonstration program, said, "The information gathered will be compiled into a larger report and delivered to each product team so that designers can incorporate customer feedback into the planning of future products. I truly believe that the information from our end users improves our products."

In addition to success for the program, a close relationship with customers also puts the program manager in a position to cultivate new business opportunities for the enterprise. This is especially true in situations where a program is executed via a contractual arrangement. Contract changes and extensions can become opportunities for additional business.

Program managers also need to understand who the competitors are and what they have to offer. To create a competitive advantage, it requires that competitors are known, that the competitor's solutions are understood, that the solution being developed by the program team can be compared to competitor's solutions, and that the differences between the solutions and the value proposition of the program team's solution can be explained to potential customers.

This exercise is often conducted by program teams but is treated as a one-time event either as part of the concept approval or the business case. However, to ensure competitive viability of the capability under development, this exercise needs to be repeated periodically throughout the life of a program—at least at every decision checkpoint (Chapter 6).

THE MASTER INTEGRATOR

Some years ago one of the authors attended an industry seminar at a highly respected university in the U.S. Before the seminar began he had a brief conversation with a person named Mary who worked for Corning (the glass manufacturer). During the conversation Mary asked what the author did for the company he was working for, to which he told her that he was a program manager. "Ah, the master integrator," was Mary's reply.

This was the first time we heard the term *master integrator* in relation to program management, but we continue to use it in our work because it accurately describes the second primary role of the program manager.

So why does a program manager have to function as a master integrator? The answer to that question is contained in Chapter 4. The program manager is responsible for providing a whole solution made up of many pieces. A program has to be decomposed into a number of projects representing smaller elements of specialized work, and then work to reintegrate the output and outcomes of the specialized work into the holistic solution.

A Focus on Integration

On occasion we have the opportunity to evaluate the current state of the program management function within an organization. One of the first things we try to do is determine if the organization is primarily project-oriented or program-oriented (Chapter 1). Observance of the roles of the firm's program managers provides a quick analysis—specifically, observing if the program managers are behaving as administrators, coordinators, or integrators. A predominantly administration or coordination role for the program manager usually indicates that the organization is project-oriented.

In project-oriented organizations, programs tend to be characterized as a loose grouping of functional specialists who operate within their area of specialty, and functional agendas and goals take precedence over a *common* vision and goal. It is also common to see a high level of dysfunction among the functional departments, and between the functions and the program office. This situation is normally an indication that the integration role of the program manager is not well understood or being performed within the business.

By contrast, in organizations where the program manager has an integration role and responsibility, it can be observed that the various functions and specialties are working collaboratively toward a common vision and goal and are doing so to create a holistic solution.

The disaggregation and reintegration of work is the centerpiece of integration and systems thinking. By performing the role of master integrator on a program, the program manager, who is focused on the whole solution, is in effect working as a systems integrator (from a programmatic perspective rather than a technological perspective). The example in Figure 10.5 illustrates this point.

The program manager, in the illustration, oversees the disaggregation of the whole solution (only a few of the elements are shown in the example), into a program architecture composed of multiple projects and enabling components. He or she then oversees the integration of work output from each of the components to create the consolidated whole solution. The true value and benefits of the program output can only be realized when the activities associated with each of the projects and enablers are integrated together into a holistic solution for the customer.

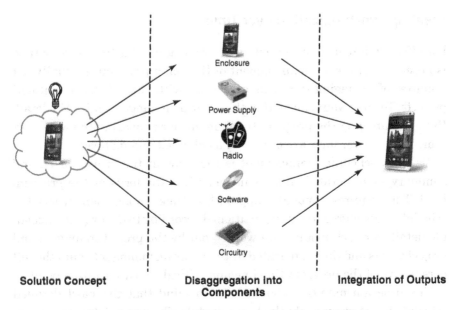

| Solution Concept | Disaggregation Into Components | Integration of Outputs |

Figure 10.5   Integration as part of systems thinking.

Managing Interdependencies

Referring to Figure 4.8 in Chapter 4, we consider the *horizontal* dimension of a program where the program manager is responsible for synchronizing the work flow of the constituent projects across the specialist functions of the firm. This is accomplished through cross-project interface definition, coordination of all program activities, driving collaborative communication on the program, and synchronizing the delivery of the project interdependencies.

While the role of each project manager on a program is to manage the creation and delivery of their project deliverables and outcomes, an important aspect of the integration role of the program manager is overseeing the hand-off of deliverables between project teams on the program. This output-input relationship creates a network of interdependencies that has to be established and managed at the program level.

Establishing the network of interdependencies is best accomplished through the practice of program mapping (Chapter 7), and the creation of a program map. The program manager can use the program map to manage the cross-project interdependencies during the execution cycles of the program.

Creating Synchronization over Time

For effective integration to occur, synchronization of activities over time is necessary. However, management of time on a program is usually not a matter of managing the tasks associated with each of the constituent projects. To do so would mean that the program manager is mired in detail. Rather, managing the program timeline involves ensuring the work outcomes and deliverables are occurring as planned and mapped.

Detailed schedule management is the focus at the project level, and summary, or integrated time management, is the focus at the program level. This requires a modular approach to time management where the schedule is disaggregated and partitioned according to the projects. Schedule details for each module are worked out by the project managers and project teams and then integrated by the program manager to gain the full perspective of the program timeline and critical synchronization points.

The program manager should keep in mind that the most detailed schedule is not necessarily the best schedule. Too much detail can divert attention to one aspect of a program—the schedule—and away from the other critical aspects of managing a program and leading a team. If the program manager focuses on the critical synchronization points occurring over time and lets the project managers focus on the detailed schedule for their respective projects, a good balance for effective timeline management is achieved.

LEADING THE PROGRAM TEAM

In program management, like any other discipline, there is a distinct difference between good program managers and great program managers. Good program managers are those that can manage the business and operational aspects of a program. Great program managers are those that have the ability to combine strong management skills with strong leadership skills. The primary differentiator between good and great is one's ability to build a high-performing program team and lead the team toward the achievement of the business objectives by establishing a vision, or end state, and then guiding the work of the entire team toward that vision. At the same time the leader helps the team to establish pride in their work as well as knowledge of how their work contributes to the program goals.[5]

The program manager uses his or her management skills to utilize the physical resources (human skills, raw materials, and technology) of the enterprise to ensure that the work of the team is well planned, is performed on schedule and within cost, is delivered with a high level of quality, and is executed productively and efficiently. However, because program managers rarely have resources reporting to them directly, they must use strong leadership skills to build relationships to influence, focus, and motivate the program team. In doing so, program managers utilize the emotional and spiritual resources of the organization to create and deliver the business benefits expected by senior management.

For the program manager, this means that he or she is responsible for keeping the team of functional specialists performing as a cohesive group to achieve the objectives of the program, while helping each project team (and individual) understand how their element contributes to the creation of the whole solution. Program managers must grapple with four essential issues while building and leading the program team: 1) establishing the team vision, goals, and objectives, 2) defining the roles and responsibilities for each member of the team, 3) instituting the team norms and work procedures, and 4) managing personal relationships.[6]

Establishing a Sense of "We" versus "Me"

Pulling a group of functional specialists into a cohesive, effective team requires the team members to perform as a single entity and identify with the program more than their functional specialty and that they remain focused on a common vision of what they are trying to achieve.

The program vision defines the end state that the program is trying to achieve. As discussed in Chapter 3, it is the whole solution and the set of business goals that the program is commissioned to achieve that is commonly referred to as "the big picture." Everything that happens on a program should do so in the context of the big picture, or program vision.[7]

One of the essentials of strong leadership is that the leader *pulls* upon the energy and talents of the team, rather than pushing, ordering, or manipulating people. Leadership is not about imposing the will of the leader, but rather about creating a compelling vision that people are willing to support and exert whatever energy is needed to realize it.[8]

Establishing the right *level* of metrics and success factors is vitally important to establish "we" versus "me" thinking for a program team—in particular, the business success factors. Each project team within a program needs to focus on their key performance indicators, but consistently communicating the business success factors for the program to the project teams focuses the entire program team on success criteria that can only be achieved by the *collective* success of the team.

The program vision, program objectives, metrics, and the business success criteria are devices that the program manager uses to pull the team together and to keep the project specialists focused on the program goals, rather than just the goals of their specific function. It is the basis of empowerment that allows the program manager to establish power of influence, rather than positional power.

Establishing Team Chemistry

A program team may consist of the top talent within an organization, but they will not reach a high level of performance without a certain bonding of spirit and purposefulness. It is this bonding, or team chemistry, that motivates team members to work together collaboratively for the common success of the team.

Of course, when a group of people form a team, their personalities may not immediately gel. Acculturation of personalities, ideas, shared values, and goal alignment takes time as well as intentional effort (sometimes considerable effort) on the part of the program manager. People from a diverse set of backgrounds and experiences will bring different behaviors, routines, values, and ideas about the work of the team. The program manager must embrace this diversity of people on his or her team as individual members that make up a collective work unit, and act as a coach and role model for the rest of the team to help them embrace the value of diversity. What the team must learn is that there is great benefit to having differences in personality, values, opinions, and ideas working toward an optimal business solution.

There are a number of things that successful program managers do to accelerate the establishment of team chemistry. These include establishing team norms, fostering social presence, and celebrating team successes. We are firm believers in the value of bringing the team together for face-to-face collaboration exchanges whenever possible. This investment in time and sometimes money pays significant dividends in establishment of team chemistry.

Establishing Team Norms

With his or her team members, a program manager can lead and facilitate a series of discussions on how the team will perform its work and conduct itself. In particular, focus on acceptable and unacceptable behaviors, meeting types and forums needed, communication preferences, how decisions will be made, and reporting methods, messaging, and frequency. Make sure to gain agreement on the norms and then set the ground rules and expectation that the team behaves and acts in a way that upholds the norms.

Foster Social Presence

To prevent some members of the team from becoming "invisible," make sure that all team members know one another and continue to foster connections. This is especially important for geographically distributed teams. Initial introductions are critical, and must go beyond the usual statement of name, functional group, and their understanding of the role one plays on the team. It's also more effective to ask each individual to describe his or her expertise and professional background, as well as something interesting and significant about them. Some program managers have created simple social networking websites for their teams that provide member profiles. The goal of the program manager here is to create a team environment. So, getting to know one another (professionally and personally) is important and necessary to achieve team cohesion.

Celebrate Success as a Team

Making sure that the entire team participates in team celebrations goes a long way toward focusing team members on team accomplishments over individual accomplishments. The celebrations do not have to be large or even formal in nature to be appropriate and appreciated by the team members. Successful program managers look for opportunities to recognize team accomplishments throughout the duration of the program at key milestones, major events throughout the calendar, and even as surprises to the team to help manage program-related pressures.

Building and Sustaining Trust

Trust within a team is the foundation of effective collaboration. For a team to reach its highest level of performance, much attention has to be paid to building trust between the team members and between the team and the program manager, sponsors, and other stakeholders.

Table 10.1   Trust creator and destroyers.

Trust Creators	Trust Destroyers
Act with integrity	Demonstrate inconsistency between words and actions
Communicate openly and honestly	Withhold information or support
Focus the team on shared goals	Put personal gain over team gain
Show respect to team members as equal partners	Engage in lies, sabotage, and scapegoating
Listen with an open mind	Listen with a closed mind

In his book *The 21 Irrefutable Laws of Leadership*, John Maxwell uses the analogy of building trust as either putting change into your pocket or paying it out.[9] Each time a program manager forms a new team, he or she begins with a certain amount of change in his or her pocket, representing the inherent trust a person receives from his or her position as the program manager. As the program progresses, the program manager either continues to accumulate change in his or her pocket by building trust across the team, or finds that the pocket begins to empty when trust is depleting.

Table 10.1 lists the factors that both destroy and create trust for a program team. Obviously, the team leader is best served by acting upon the trust creators and avoiding the trust destroyers.

Building strong relationships between team members is also an important factor in enhancing and sustaining trust on a program team, especially later in the life of the team. In many ways, program managers are the glue that holds teams together. Establishing trust in themselves and between team members based on demonstrated trustworthiness can be a difficult task. They should begin by setting the expectation that the teams perform within the confines of the elements that demonstrate one is worthy of mutual trust:

- Perform competently
- Act with integrity
- Follow through on commitments
- Display concern for the well-being of others
- Behave consistently

The program manager must treat trust as his or her greatest asset and realize that it is important to consistently model the behaviors that exemplify competence, connection, and character in leading the team. The best program managers we have encountered realize the need to go beyond establishing expectations that the team demonstrate the trust building behaviors stated above. They also model the behaviors on a daily basis by always standing behind and supporting their teams, never demonstrating favoritism, and accepting full accountability for the actions and results of their teams.

Making Tough Decisions

There can be thousands of decisions, large and small, that a program manager will encounter during the course of a program. To prevent even a small number of these decisions from being barriers to progress, a program manager needs to be proficient in collecting all necessary facts, analyzing the pertinent data, and then driving to a decision. Nothing can be more frustrating and paralyzing to a team than waiting for a decision that prevents forward progress.[10]

The ability of the program manager to make timely decisions will be considerably influenced by their degree of experience and their ability to think of the program from a holistic perspective. By a holistic perspective, we mean that program decisions need to be aligned with the objectives of the program and the strategic goals of the business. Losing sight of the strategic reasons for a program while making a series of large and small decisions is a primary reason why some programs become misaligned with the strategic goals of the business and ultimately fail.

As a safeguard, the program manager must establish boundary conditions that serve as guard rails to prevent goal misalignment. The more concisely and clearly the boundary conditions for a decision are stated, the greater the likelihood that the decision will be effective in accomplishing the direction that is needed and ensuring that the direction is consistent with the business goals driving the need for the program. The program strike zone (Chapter 9) is an important decision support tool that helps to keep program decisions in alignment with the business goals.

Empowering the Program Team

With leadership, comes power. To lead, one must have a relationship with people who are willing to follow. For the program manager, those people are the IPT core team specifically and more generally the entire program team. Every leader's potential is determined by the abilities and actions of the people who are closest to him or her. If the leader's inner circle of people are strong and capable, the leader can make a big impact within an organization.[11] This is true for the program manager whose inner circle consists of the members of the core team.

As discussed in Chapter 5, careful selection of the IPT core team members is critically important for this reason. First and foremost, program managers should select core team members who are experts in the function that they represent within the organization. The second critical criteria for core team selection is a person's ability to step outside of their area of expertise to effectively collaborate with the other members of the team toward the achievement of the program vision. One of the most significant ingredients for program success is the cooperation and collaboration between the project teams within the program, in which everyone understands that they cannot succeed unless everyone else succeeds.[12]

Empowerment is the sharing of power from one person to another and granting them influence and authority to take responsibility and make decisions within their sphere of work. The program manager empowers the core team by giving away a portion of his or her own power to the project managers and functional representatives on the core team. Thus, the project manager and functional representative will succeed by exercising control over their own decisions and resources. This is an important concept for a program manager to grasp. Nothing is more disempowering than giving the core team a lot of responsibility for program success, then not empowering them to work autonomously. Effective leaders get the most out of the members of their inner circle by treating them in a way that bolsters their self-confidence and provides them with the necessary resources to succeed.[13] A strong sense of self-confidence makes it possible for the IPT core team members to take control of their portion of the program, which allows the program manager to focus on the big picture.

Establishing an environment of "no fear" is also a critical element in developing a team's confidence. As team members are granted greater empowerment, they will begin to act more on their own and rely less on the direction of the program manager. They will take on a greater sense of responsibility for their work output, become more comfortable with making decisions and solving problems on their own, begin to act proactively instead of reacting to change, and ultimately become more motivated to succeed.

As a senior program manager for a major aerospace company told us, "A no-fear environment is needed so people don't have to fear coming to the program manager with the real answer, especially if it's not a pleasant answer. You have to let people stumble a bit in order for them to learn and gain confidence."

The section titled "The Leadership Story of a World-Class Program Manager" provides an example of personal leadership principles one practicing program manager has established for himself through his experiences in leading many program teams.

The Leadership Story of a World-Class Program Manager

Pali Pafgan is a senior program manager for a marketing company who has spent considerable effort and time honing his team leadership abilities. Pali shared with us the personal philosophy and principles he has developed over time for leading his program teams.

Principle 1: *Obtain commitment.* Pali defines the role of each stakeholder, emphasizing that they are equally important. He tries to get all primary stakeholders involved in his program as equal partners who build consensus whenever needed.

Principle 2: *Insist on transparency.* Pali shows an honest interest in others' opinions, and insists that everybody explicitly state his or her interest in the program. He works hard to obtain information inside and outside the company and share it in a facilitated and informed decision-making model.

Principle 3: *Empower the team.* Pali creates opportunities for his team members to hone their talents. He also insists that the team develops its goals and work plan and holds them accountable for their own success.

Principle 4: *Focus on human relations.* Pali concentrates on relationships within the team, trying whenever possible to select members based on their functional and people skills. His intent is to provide team building training to members who needed it and define the team as the only place for decision making.

Principle 5: *Promote learning.* Pali encourages members to behave like entrepreneurs and take risks. He also wants members to feel safe in the program environment and to share their agendas and opinions for all members to scrutinize.

ENDNOTES

1. Stagliano, A. R. "Cash Is the Lifeblood of Every Contractor." *Building Profits*, 2011.

2. Keogh, Jim, Avraham Shtub, Jonathan F. Bard, and Shlomo Globerson. *Project Planning and Implementation*. Needham Heights, Mass.: Pearson Custom Publishing, 2000.

3. Demarco, Thomas, and Timothy Lister. *Waltzing with Bears: Managing Risk on Software Projects*. New York: Dorset House Publishing, 2003.

4. "HP Gets Closer to Users by Design," *Oregonian*, January 6, 2006.

5. Bennis, Warren G., and Burt Nanus. *Leaders: Strategies for Taking Charge*, 2d ed. New York: HarperCollins Publishers, 1997.

6. Lewis, James P. *Fundamentals of Project Management*. New York: AMACOM, 1997.

7. Maxwell, John C. *The 21 Irrefutable Laws of Leadership*. Nashville, Tenn.: Thomas Nelson, 1991, p. 58.

8. Kouzes, James M., and Barry Z. Posner, *The Leadership Challenge*, 3d ed. San Francisco: Jossey-Bass, 2003, p. 143.

9. Maxwell, *The 21 Irrefutable Laws*.

10. Kouzes and Posner, *The Leadership Challenge*.

11. Maxwell, *The 21 Irrefutable Laws*.

12. Kouzes, James M., and Barry Z. Posner. *The Leadership Challenge*, 3d ed. San Francisco: Jossey-Bass, 2003, p. 187.

13. Martinelli, Russ; Rahschulte, Tim; and Waddell, Jim. *Leading Global Project Teams: The New Leadership Challenge*. Toronto: Multi-Media Publications, 2007.

Chapter 11

Program Manager Competencies

What was your motivation for choosing program management for developing products? We posed that question to Gary Rosen, corporate vice president for Applied Materials Corporation. Rosen responded by stating the following:

> When observing the differences between poorly run product development efforts and well-run efforts, I noticed the difference was that the well-run programs had a true program manager in charge. These people had a broad skill base that is needed—good people skills, good business acumen, and good system skills. Unfortunately, not a lot of people have these broad skills.

Given the roles and responsibilities of the program manager discussed in Chapter 10, it is a rare program manager that comes to these roles totally qualified to meet all the skills and competencies required. The successful program manager is constantly seeking to learn and broaden his or her knowledge and experience in order to take on more complex and critical programs. Senior management in turn needs to create a positive learning environment to encourage program managers to continually seek improvement and growth.

We developed the program management competency model in order to address the breadth, depth, and complexity of the program management role. The model provides senior management with an excellent aid to assist an organization in developing program management as a true discipline that can provide value through increasing the likelihood of successfully achieving intended business results. The more experienced and capable program managers are, the greater the probability that they will successfully deliver the business goals of the firm.

In this chapter we use the program management competency model to detail the knowledge, skills, and abilities needed for program managers to continually grow as professionals and consistently succeed in their role. Additionally, we discuss the key organizational enablers needed to make the competency model fully effective and to adequately support the program management discipline within an organization.

THE PROGRAM MANAGEMENT COMPETENCY MODEL

The program management competency model has been designed to provide the knowledge, skills, and abilities to systematically support the recruiting, staffing, professional development, and career planning of the program manager (see "What Is Competence?"). The information presented in this section has been derived from companies that use program management as a true discipline and business function. Although the technical aspects of developing a new capability or transforming an organization are critically important, much of the success of a program is behavioral and human-oriented in nature.

A vice president and general manager for a major U.S.-based defense contractor underscored this assertion when describing to us the critical skills needed for a program manager:

> A couple years ago I would have started this discussion with the need for a program manager to have strong technical skills. Now that I truly understand the role of the program manager I wouldn't include technical skills. More important is a program manager's sense of maturity in dealing with people who have differing expectations—the customer, senior managers, and the program team.

What Is Competence?

The word *competence* means the ability to do something well. With respect to management, we found that the term is highly overdefined. Take, for example, the following definitions:[1]

"An underlying characteristic of a person in that it may be a motive, trait, skill, aspect of one's self-image (or social role), or a body of knowledge that he or she uses." While comprehensive, this definition may not be very user-friendly or practical.

"The characteristics of a manager that leads to the demonstration of skills and abilities and results in effective performance within an occupational area." We find this definition more practical in nature, but it doesn't address the fact that competent managers need a strong knowledge base to effectively apply their skills and abilities.

"The knowledge, skills, and qualities of effective managers used to effectively perform the functions associated with management in the work situation." We feel this definition best describes the competencies needed by practicing program managers.

While there are many more definitions, we feel the final definition above best describes the competencies needed by program managers to effectively manage complex and demanding programs in the contemporary business environment. A simple algorithm sums it up best:

Competence = Knowledge + Skills + Personal Qualities + Experience

The objective of the program management competency model is to provide a set of competency criteria that can be used for the continued development and career growth of program managers. The four competency areas in Figure 11.1 are mutually dependent on one another for achieving overall performance success. It is the interworking of all of these competencies together that drive the synergy of the model. It is also important to note that the competencies discussed apply to program managers who lead either co-located teams or geographically distributed teams.

Many program managers come to the role by way of having a successful track record as a project manager. One aspect from the summary view of the core competency model is that project management is but one element of the program management discipline. This underscores the discussion in Chapter 1 that showed the distinction between program management and project management. One can visualize from Figure 11.1 that program management encompasses a broader role than project management. In addition to strong project management skills, the successful program

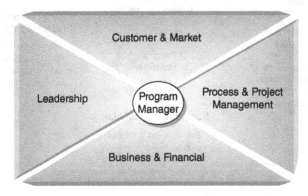

Figure 11.1   Program management competency areas.

manager needs to gain proficiency in broad-based leadership, business and financial, customer and market, and other process competencies. Without this added proficiency, the transition from a project management role to a program management role is many times frustrating for the individual, the organization, and the customer.

When detail is added to each of the competency areas, the core skills of the program manager emerge. The following sections provide that detail for each of the four program management competency areas.

CUSTOMER AND MARKET COMPETENCIES

Customer and market competency involves having a full understanding of the markets a company serves and how an organization's capabilities are being utilized by the customer and end user. It should be noted that we use the term *market* to describe either the external or internal market. The better the program manager and his or her team can closely align the capability under development or the organizational transition end state with the customer's needs, the more it will enhance the potential for customer satisfaction and the successful achievement of the business results intended. Critical skills in which the program manager should gain proficiency within the customer and market core competency are shown in Figure 11.2.

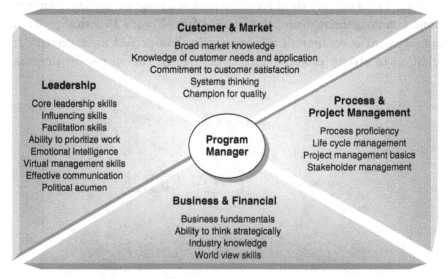

Figure 11.2   Critical program management skills.

Customer and market competency means that the program manager has a comprehensive understanding of the market or organization into which the program output will be deployed, the types of customers, the competitors, and the general attributes and trends in the market. More broadly speaking, the program manager should possess a working knowledge of the key technologies that a firm's solutions are based upon, as well as the future trends within their industry.

New capabilities and solutions are successful when they have been designed to directly meet the application needs of the customer base. Today it is common to have key customers participate in the development process. Therefore, the program manager should be knowledgeable and sufficiently skilled to incorporate the valuable input of the customer into the program management process—what some call *customer intimacy*. Customer intimacy is all about understanding what the customer really wants and needs. This intimate knowledge of your customer is critical to maintaining a good working relationship with your key customers and end users (see "Do You Know What Your Customers Really Want?").

Do You Know What Your Customers Really Want?

Program engineers at Accuracy Inc. were proud that they practiced the concept of customer intimacy, and, as they put it, "It showed in their practice." They saw customer intimacy as the ability of their product development teams to recognize, internalize, and build the customer's needs into their customized interconnecting cable products. To check how much they were "customer intimate," they were advised to design and administer a survey that would ask 36 customers how satisfied they were with Accuracy's performance. The survey identified 15 dimensions of customer intimacy, which included, for example, the quality of the joint product definition process, quality of the rapid prototyping development process, ability to manage product changes, adherence to program milestones, and adherence to the final delivery date. For ratings, the survey used the following scale: 1-poor, 2-fair, 3-good, 4-very good, and 5-great.

On 2 of the 15 dimensions, Accuracy Inc. received very good ratings or higher for adherence to the final delivery date and program milestones. The remaining 13 dimensions revealed average ratings below 3, of which most dimensions were rated about 2, or fair. As he looked at these ratings, the director of product development wondered if his group was really "customer intimate." What do you think?

In program management, delivering the whole solution (Chapter 4) is a primary means to achieving customer satisfaction. The program manager needs to be able to demonstrate a commitment to the customer and demonstrate knowledge of customer application and needs. Those skilled in systems thinking can view projects and activities from a broad perspective that includes seeing overall characteristics and patterns rather than just individual elements. By focusing on the entirety of the program, or in essence the system aspects of the program (inputs, outputs, and interrelationships), the program manager improves the probability of delivering the whole solution and meeting the expectations of the customer. This involves the ability to see the big picture, crossing boundaries, and being able to combine disparate elements into a holistic entity. Usually this ability resides in people with very diverse backgrounds, multi-disciplined minds, and a broad spectrum of experiences.[2]

The program manager, as the quality champion for his or her program's customers, needs to ensure the program results meet or exceed the quality expectations of the customer. The program manager should possess a bias for action, be able to think globally, and assure that quality, reliability, manufacturability, serviceability, and regulatory compliance objectives are achieved.

BUSINESS AND FINANCIAL COMPETENCIES

To be successful from a business perspective, the program manager must possess sufficient business skills to understand the organization's business model and financial goals. This requires that a program manager has the ability to develop a comprehensive program business case that supports the company's objectives and strategies, the ability to manage the program within the business aspects of the company, and the ability to understand and analyze the related financial measures pertaining to the program. The areas in which the program manager should gain proficiency in order to effectively manage the business of a program are shown in Figure 11.2.

When operating in the role of the business manager proxy, the program manager must possess the core business acumen needed to manage the business aspects of a program as detailed in Chapter 10:

- Aligning a program to business strategy
- Creating and managing to the business case
- Managing business risk
- Monitoring customers and the market

- Managing intellectual property
- Managing program resources

Other business fundamentals include capabilities in financial analysis and accounting, international management, economics, law and ethics, resource management, negotiation, and communication. This implies that there is also a solid understanding regarding the integration of these elements and related impact on one another.

The program manager is required to think strategically in order to align the program and its constituent projects to the strategic business goals of the organization. This includes understanding how a firm or organization performs strategic planning and its relationship to portfolio planning and be able to separate aspects of strategic thinking from tactical and operational elements as the need arises. A part of strategic thinking involves a basic understanding of the industry in which a business operates and how the firm's strategy fits with the direction of the industry long term.

Industry trends, knowledge about competitors, and supply chain implications are a fundamental part of keeping a program viable from a business perspective. The program manager should possess a strong business sense and integrative capability to successfully manage the broad array of strategic, business, and financial attributes related to his or her specific program and how these pertain to the industry the organization serves.

As previously mentioned, much of the work we do today is performed across national boundaries. Therefore, the program manager must possess a world view of his or her role and be knowledgeable and understand the global and local site environment including the social, political, and economic trends in order to be successful. This should also include a specific working knowledge of the local site and international economics and business requirements under which the program is operating. The world view capability is significantly enhanced if a program manager has had earlier experience and exposure to various international cultures and markets through earlier work assignments or extended visits.

PROCESS AND PROJECT MANAGEMENT COMPETENCIES

A program manager should be well versed in core organizational processes and project management competencies, and be able to abstract the competences for use at the program management level. The program manager needs to be competent in processes to handle program level issues. Also, both the program manager and the project managers on the program must possess operational competencies, including project management methods and tools, to effectively manage the operational elements of the program. The areas in which the program manager should gain process and project management proficiency are shown in Figure 11.2.

An important aspect of this core skill set is that of becoming proficient at possessing a solid working knowledge of the specific processes and practices of the company, knowing how things get done, the policies and procedures that must be adhered to, and who must be involved to approve various aspects of the program. If it is a product-based company, for example, the program manager must be thoroughly familiar with the firm's new development processes to ensure that the program adheres to the company's requirements and expectations as to how products are designed and built. The program manager that develops this competency will increase his or her probability of gaining team members' confidence and trust. They will become confident that their leader knows how to get things done in a timely and successful manner.

A challenge in leading a program team becomes one of ensuring that all members of the team are consistently following the same processes and methods, and are using the same tools when appropriate. A foundational element in driving process consistency is to ensure that a common life cycle and decision framework (Chapter 6) is being used by all. This will help to drive common language and terminology, establish a common cadence of activities, and provide common decision and synchronization points throughout the program cycle.

Project management skills are primarily determined by the particular methodology or methodologies used on a program (PMI's *PMBOK®*, PRINCE2, or Agile for example). The program manager must be able to apply the methodology and at times be able to apply multiple methodologies at the program level. Each methodology provides key information and understanding that both project managers and program managers should master in order to possess the highest probability of success in achieving the objectives and expectations of their customers and management. When it comes to project management competencies, the program manager role has many similarities to that of the project manager with the key added dimension of leading and integrating the work being done by multiple projects simultaneously.

Stakeholder management skills are critical for program manager success. The program manager first must know how to determine the organizational landscape in which the program is to operate. He or she will likely have many stakeholders both internal and external to the organization who need to be influenced (see "Even a Salad Vendor May Be a Stakeholder"). Good stakeholder management first of all involves understanding who the program stakeholders are and their needs, understanding the level of influence each stakeholder has on the program, and understanding their allegiance and attitude toward the program (never assume all stakeholders want the program to succeed). From this information, the program manager can determine which stakeholders he or she needs to manage and develop a stakeholder influencing strategy (Chapter 7). The program manager must then be able to use and extend this knowledge to develop the ability to choose the right mode of communication to address customers, senior management, team members, suppliers, and others. This involves knowing when to see people face to face, when to send messages, and when to avoid them altogether.

Even a Salad Vendor May Be a Stakeholder

This is a bizarre example of what can happen if a comprehensive stakeholder analysis is not conducted on a program. The program involved a fast-track transfer of manufacturing technology from Europe to a Middle East country. Just after the beginning of the program, the program manager got a call from a vendor claiming that all roads leading to his office were blocked by groups of violent people. As a result, some important computer equipment couldn't be delivered. A quick check proved this call correct. The people were local butterhead salad farmers who were unhappy that the foreign contractor was importing salad from Europe, rather than buying it from them. The siege went on for days and the delivery delays were impacting the program schedule. Finally, the program manager figured out his mistake—one cannot ignore the relationship with local communities that have a big stake in the program. In this case the salad farmers were part of the local community, and therefore stakeholders in the program. Once the oversight was discovered, the contractor began buying local salad and program deliveries proceeded without delays.

LEADERSHIP COMPETENCIES

Leadership competencies are how we describe the "people skills" program managers need to be successful in leading a team of people. Leadership competencies involve those that are needed to effectively lead multiple cross-disciplined project teams that are a part of the program. This pertains to both co-located teams as well as in highly distributed or virtual teams. Specifically, we are referring to organizational leadership competencies. The areas in which the program manager should gain leadership proficiency in order to effectively lead a program team are shown in Figure 11.2.

Core Leadership Skills

First and foremost, a program manager needs to have the capability to build, coalesce, and champion the team to deliver solutions that will satisfy the company's goals and customer's needs. The foundational elements of effective team leadership apply whether one is leading a domestic team that is co-located in a single site or a global team that is distributed across multiple countries. Success begins with the core principles of team leadership, as discussed in Chapter 10:

- Creating a common purpose
- Establishing team chemistry
- Building and sustaining trust
- Demonstrating personal integrity
- Empowering the team
- Driving participation, collaboration, and integration
- Communicating effectively
- Managing team conflict
- Making tough decisions
- Providing recognition and rewards

Influencing

In today's organizations, members of a program team rarely report directly to the program manager. This requires the program manager to become proficient in influencing the actions of the team and stakeholders. Influencing traits of a strong program manager include being socially adept in interacting with others in any given situation, having the ability to assess all aspects of information and behavior without passing judgment or injecting bias, and being able to effectively communicate your point of view to change opinion or change course of action. As John Maxwell states in *The 21 Irrefutable Laws of Leadership*, "Leadership is influence—nothing more, nothing less. If you don't have influence, you will *never* be able to lead others."[3]

Facilitation

Simply put, facilitation is the act of assisting team members to reach their collective goals by helping to make team communication and collaboration easier and more effective. Good facilitation skills help to ensure that relationships between team members continue to develop and that ongoing communication and collaboration between team members is occurring as needed—productively. Core facilitation skills include the ability to draw out varying opinions and viewpoints among team members, to create a discussion and collaboration framework consisting of a clear end state and decision and collaboration boundaries, and to summarize and synthesize details into useful information and strategy. Other beneficial facilitation skills include using personal energy to maintain forward momentum, being able to rationalize cause and effect, helping team members to establish one-on-one relationships, and keeping team members focused on the primary topics of discussion and collaboration.

Since a significant amount of communication and collaboration between team members occurs in team meetings, the program manager has to develop skills in planning and conducting effective meetings. Team meetings will vary from face-to-face meetings phone conferences video conferences, Internet-based data sharing meetings, or some combination of all of these. This involves preplanning an agenda with time-boxed topics, sending any materials that will be used in the meeting to all members prior to the meeting, setting the meeting ground rules, facilitating the discussion appropriately to ensure a mutual understanding of all conversations, and periodically checking to see if quiet members are understanding the discussion and are fully engaged.

Prioritization of Work

Effective program managers are continually prioritizing their work and the work of their team to achieve the greatest return from the actions of the team. There is an enormous opportunity for the program team to spend too much time on work that is of little value due to the complexity and scale of most programs. Prioritization of work begins with validation of the core assumptions driving the direction of the program with the primary stakeholders and program governance body. If the assumptions behind the priorities are incorrect, it is quite possible that the priorities themselves will be incorrect.[4] For example, if cost containment is the highest priority of the program, then the program manager must be emphatic about staying within the financial constraints. If technological leadership is the highest priority, the program manager will need to keep the team focused on the technical aspects of the program. The ability of the program manager to focus the work of the project teams on the highest priority needs of the program is crucial for maintaining a high level of team priority.

Emotional Intelligence

Daniel Goleman described emotional intelligence as "managing with heart."[5] Emotional intelligence skills involve being acutely in tune and sensitive to emotional responses of program team members. Emotional intelligence skills consist of two types of competencies: personal competence and social competence. Personal competence involves both self-awareness and self-management, where self-awareness is one's ability to accurately perceive one's own emotions and moods in the moment and understand one's tendencies in various situations. Self-management is the ability to use awareness of emotions to stay flexible and direct one's behavior positively. Thus, self-awareness involves staying on top of one's reactions to team members and others and managing one's own emotional self-regulation to think before one acts or reacts. Social competence includes social awareness and relationship management skills that drive at understanding others and managing relationships.

Social awareness is one's ability to accurately pick up on the emotions of others and to understand what is really going on with them whether one agrees with them or not. In essence, it is applying empathy and appropriately understanding and reacting to the emotional needs of others.

Virtual Team Leadership

Today's business models have created additional team building challenges for program managers (see "Virtual, Discontinuous, and Condensed Team Building"). It is common for program team members to be distributed across multiple countries. Skills for managing virtual teams (non-co-located teams) have become an emerging critical skill for program managers.[6] There are many aspects to successfully leading a geographically distributed (virtual) team. We will cover a few of the most critical.

Virtual, Discontinuous, and Condensed Team Building

True program leaders forge the vision for the team, evangelically communicate it, and inspire and motivate team members to follow them. All this is typically associated with leading a single program at a time in a co-located manner. But business models of today have brought new leadership challenges for program managers, including leadership of virtual programs and simultaneous leadership of multiple programs.

The first challenge, leading virtual program teams, raises the following question: Is it possible to build true leadership in virtual teams when members are geographically, culturally, organizationally, and time zone dispersed? Patrick Little, a senior IT program manager for a leading research hospital, answered that question this way:

> *Leading a virtual team is possible, but it takes additional effort from all members of the team. Our virtual program team has an assigned program leader, me, but leadership roles are shared within the team. Leadership responsibilities include motivating, information- and opinion-seeking, mediation, facilitating communication, removing barriers, lubricating interfaces, and making each conflict functional so it could be used to improve the quality of our decisions. All these aspects are important for building team spirit. However, doing this in a distributed team structure requires a lot of travel on my part. Frankly, I am on the road all of the time.*

The second challenge, leading multiple simultaneous programs, is equally difficult. In a classic situation of leading a single program at a time, program managers can focus on leading a single team without needing to split their time and effort between multiple program teams. Because program managers managing multiple programs need to build multiple teams concurrently, their time for each team is limited. Consequently, they need to apply a condensed, fast team-building method. Similarly, they are expected to operate in a discontinuous manner—lead one team, discontinue, and then lead another team.

As business models continue to change, so do program management practices and competencies. And as managing programs become more complex, we find that program management core competencies expand to adjust to the complexity to continually provide improved business results.

Many virtual teams are multi-cultural entities due to geographic distribution. Competence at cross-cultural management is critical in leading teams in a geographically distributed environment. Cross-cultural management includes awareness of cultures you are directly involved in and understanding attitudes, differences, and behaviors. Its focus is toward improving the interaction and working relationship between team members, management, and suppliers from all the cultures represented in the direct and broader team. It requires program managers to examine their own biases and prejudices, develop cross-cultural skills, and when possible, observe and learn from culturally proficient role models. Only through understanding and appreciating the unique characteristics of their culture can the program manager show the proper level of respect and understanding that each team member deserves.

Effective Communication

Communicating in a virtual environment requires us to broaden our perspective and appreciation for the entire communication process due to the comprehensive set of challenges facing the exchange of meaning. Communication is any behavior another person perceives and interprets as the understanding of what was meant. It includes sending both verbal messages (words) and nonverbal messages (tone of voice, facial expressions, behaviors). Communication therefore involves a complex multi-layered and dynamic process through which we exchange meaning.[7]

Communication on a virtual team is complicated by the physical separation of team members and the resulting reliance upon technologies to facilitate team communication. The global team leader needs to develop skills in selecting the appropriate communication technologies given the tasks required, technical competence of the team members, and infrastructure capabilities within the geographies where team members reside.[8]

Effective communication requires the ability to speak multiple disciplinary languages—business language when communicating with senior management, user language when communicating with the customer, technology language when communicating with technologists, and so on. Effective communication skills also mean that the program manager should be able to actively listen and provide clarity in difficult situations, many times serving as the translator in multidisciplinary discussions.

Rick Nardizzi, a senior program manager, told us a story of why the second program he managed was a major failure (his first program was very successful):

> Because of the success I had with my first program, I thought I knew all the answers on the second program and stopped listening to what was being said to me by the program team. I listened to what I wanted to hear

versus what they were really telling me. I wasn't looking for problems, even though they were being communicated to me by my team. Eventually, the problems became insurmountable and I was removed from the program manager position.

Political Acumen

Finally, it is important that the program manager possess both a keen understanding of the organization and have political savvy to effectively leverage and influence the power base of the company. Company politics are a natural part of any organization, and the program manager should understand that politics is a behavioral aspect of program management that he or she must contend with in order to succeed. The key is to not be naïve and to understand that not every program stakeholder sees great value in the program. A program manager must be politically sensible by being sensitive to the interests of the most powerful stakeholders, and at the same time, demonstrate good judgment by acting with integrity. The program manager must actively manage the politics surrounding his or her program to protect against negative effects of political maneuvering on the part of stakeholders and to exploit politically advantageous situations. In order to do this, it is important that the program manager possess both a keen understanding of the organization and the political savvy necessary to build strong relationships to effectively leverage and influence the power base of the company.

An effective method for working within the political environment of an organization is to leverage the powerful members of one's network. One's ability to network successfully across hierarchical and organizational boundaries is a tremendously useful skill. Networking involves becoming skilled in decision making by knowing how to determine the right parties to be involved in program decisions, understanding the impacts of a decision, and developing the appropriate messaging for the decision at hand. From a team perspective, effective networking skills give the program manager the ability to create a sense of urgency in team members who are potentially isolated from the rest of the team or are being pulled toward other competing priorities.

ALIGNING SKILLS TO ORGANIZATIONAL NEED

Throughout this book we use the program management continuum as a model to demonstrate the various ways in which program management is implemented in organizations. In Figure 10.1 the continuum was used to show how the roles of the program manager will vary depending upon how program management is implemented within an organization. In turn, the skills and competencies of a company's program managers need to align with how program management is implemented and the roles they are expected to perform.

If, for example, a firm uses program management as an administrative or facilitative function, the program managers will operate in primarily an administrative and coordination manner. The skills required for this role do not include all four competency areas of the program management competency model. Process and project management skills will be required along with some core leadership skills.

However, if this same firm wants to transform their program management discipline into an integration or business management function, their program managers will need to fulfill significantly different roles which require additional skills and competencies. Figure 11.3 illustrates this need. As the responsibilities of a program manager expand, so too must their skills and competencies. It is important to note that the skills are cumulative as one transitions to the right of the program management continuum.

Richard Vander Meere, Vice President of Global Program Management, described the top four skills needed for the program managers at Frog (a product strategy and design consultancy firm):

1. Emotional intelligence: Being able to read people and how they are reacting to a given situation
2. Project basics: Developing program scope, budget, and schedule and then managing to them
3. Savvy: Hearing, listening, and being able to quickly decide when to jump in and when to sit back
4. Influencing: Having the ability to step up in a room full of strong personalities and have a voice that affects the outcome of a situation

According to Richard, how program managers present themselves is also key:

It is important to be able to deliver a difficult message and get results without damaging your relationships.

As the leaders of an organization staff their program management positions, they must do so by being mindful of the type of skills and

	Operational Support Role		Leadership & Business Role	
Customer & Market	None	Minimal • Customer coordination • Quality assurance	Basic • System thinking • Customer needs assessment • Customer satisfaction	Significant • Broad market knowledge • Systems thinking • Customer needs assessment • Customer satisfaction
Business & Financial	None	Minimal • Budget tracking • Expenditure tracking • Earned value	Basic • Overall cost performance • ROI evaluation • Business case creation • Earned value	Significant • Business fundamentals • Strategic Thinking • Broad industry knowledge • Manage with a world view
Process & Project Management	Minimal • Process knowledge • Team coordination • Project management basics • Technical knowledge	Basic • Process knowledge • Project management acumen • Technical knowledge	Significant • Process proficiency • Project management proficiency • Integration proficiency • Stakeholder influencing	Basic • Process proficiency • Project management basics • Integration proficiency • Stakeholder proficiency
Leadership	None	Minimal • Basic team building • Facilitation proficiency • Influencing ability • Communication basics	Basic • Core leadershio acumen • Influencing proficiency • Communication proficiency • Virtual management ability	Significant • Core leadership proficiency • Prioritization proficiency • Virtual management acumen • Political savvy

Figure 11.3 Expanding skills and competencies for expanded roles.

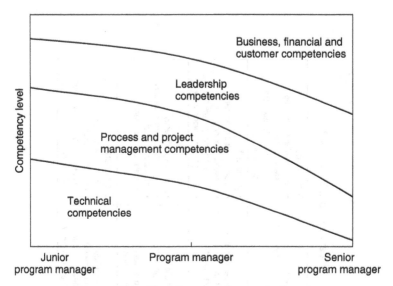

Figure 11.4   The competency—maturity mix.

experience required based upon how their program management discipline is being implemented. For example, as demonstrated in Figure 11.4, junior program managers typically have higher levels of technical competency and lower levels of business and leadership competency. This is mainly because junior program managers are commonly promoted from a specialty function or the project management ranks to lead a program. They have not yet had the time or experience leading larger, more business significant programs to hone their business and leadership competencies.

In contrast, senior program managers who lead larger and more complex programs have typically not been involved in solving specific technical problems for a period of time; therefore, they are unable to maintain their technical competency. Rather, long exposure to business, financial, and customer affairs help the managers polish these competencies.

Generally, direct technical skills become less important as the program manager takes on larger, more complex programs. He or she then delegates management of the technical aspects of the program to the project managers but may intervene on a technical issue if it becomes a barrier to the success of the program. The need for business, financial, market, and leadership competencies grows as program managers face more complex and strategically important programs.

Process and project management competencies are somewhat different. Junior program managers have modest competencies of this type because they normally have project management experience within the company prior to taking on program management assignments. As their maturity and experience grows in the program management discipline, process and project management competencies become less significant, and more focus is applied to the business management and team leadership aspects of program management.

PROFESSIONAL DEVELOPMENT

The knowledge, skills, and abilities described as part of the program management competency model are most useful for growing and developing the firm's program managers, once they are hired. This is important because future gains are made through a model of continual improvement in performance. Of course, much of what the program managers will learn will come from on-the-job training while running their programs. However, in order to further broaden their capabilities and achieve the potential of the individual, it is helpful if an organization invests in continual improvement in their discipline.

Many program managers are in the discipline because they enjoy the role and want to make it their career. Most highly effective program managers are self-motivated and demonstrate a desire to self-assess, take on a philosophy of continued improvement in their growth and responsibility, and are persistent in attainment of their personal objectives. In doing so, they may pursue growth and advancement opportunities within their company, which is important for the long-term viability of the firm.

As stated earlier, it is rare that program managers come in to the role fully proficient in all competency areas. The program management competency model has been used by companies as the guideline for continual career development of their program managers. Ongoing dialogue between the program manager and his or her direct manager should be focused (in part) on understanding the program manager's growth and career aspirations and balancing that with management's short- and long-term performance expectations of the individual. The results from this exercise serve as the basis for an individual development plan over a given period of time for the program manager (annually in most cases). It is a process of targeting where the individual currently is in their performance and capability, where they want to be in their career at some time in the future, and then developing a plan on how to get there. The specifics of the plan may include some or all of the following:

Internal training courses. Some firms offer their own training modules for the program manager. These may be web-based and available to the entire population of the company regardless of location.

External training seminars and courses. Many professional organizations and consulting firms offer a variety of courses. However, good courses focused on program management are few and difficult to find. (See www.programmanagement-academy.com).

College-degree programs. Undergraduate and graduate programs at some colleges and universities offer selected course material that can be very useful to practicing program managers.

Industry certifications. Professional organizations such as the Project Management Institute are now offering knowledge-based certifications in program management.

On-the-job training and experience. Hands-on experience is one of the most effective techniques for broadening the capability of a program manager.

Mentoring relationships. Newer program managers can gain much knowledge and capability enhancement by being mentored by an experienced program manager.

It is essential that managers give attention to issues of career paths, education and training, and in general, the development of human resources in the organization.[9] If a company does not provide an established career path for the program manager, then a program manager's best chance for advancement within his or her discipline may be pursuing more senior-level positions in other companies. For companies that have a program management office, program manager career development should be a core responsibility of the office.

ORGANIZATIONAL ENABLERS

No amount of program management knowledge, skill, experience, or training can compensate for serious organizational barriers that impede program success. Even the most skilled and knowledgeable program managers will likely fail unless the barriers and impediments are appropriately addressed within an enterprise.

It is worth repeating that senior managers of an organization play a critical role in enabling their program managers to consistently succeed. Organizational enablers are things that create or proactively encourage a positive environment to provide the maximum opportunity for success, learning, and growth to occur in program management. Enablers range from environmental factors to organizational and managerial culture, philosophy, and actions. Examples of enablers would include clear roles and responsibilities defined for those involved in the program management discipline and stable systems that can accurately and consistently measure program status and results. Other examples of organizational enablers include the following:

Enabler 1: Structure

The organization needs to be structured appropriately to enhance cross-organization collaboration. Additionally, to facilitate alignment of business strategy and execution output, a firm should be structured so that senior management and a program manager have direct access to each other to work together as a leadership team.

Enabler 2: A Systematic Management Model

This enabler involves the adoption by senior management of an integrated management system (Chapter 3) that aligns execution output directly to business strategies and goals. The model should be sufficiently broad based to enable focus on a holistic solution for the customer or client, and all elements are synchronized over the program cycle.

Enabler 3: Empowerment

Program managers need to be formally empowered by senior management with the appropriate level of authority and decision responsibility pertinent to their role as program managers. Transfer of power further enables program managers to transfer appropriate decision-making capability to team members at the local level, as appropriate, where the best information and knowledge about the program may exist.

Enabler 4: Processes

A set of formal processes and procedures with appropriate decision checkpoints along with other pertinent company policies should be established and consistently applied. This will ensure that cross-discipline work effort is in compliance with all requirements and provides an effective means for mitigating business and operational risk.

Enabler 5: Tools

Appropriate electronic communication and collaboration tools need to be in place and available for virtual communication and collaboration. These tools need to be supported by consistent procedures and team norms in order to optimize communication and the flow of information across all sites and team members.

Enabler 6: Formal Escalation Process

Processes need to be established to elevate barriers and issues outside the control of the program manager to senior management in order to get timely resolution. This also involves a system of documented success criteria, targets, and measures to indicate when escalation is needed and when it is not.

Enabler 7: Career Development

The level of responsibility and accountability granted to the program manager requires a skill set that is much more comprehensive than the traditional project manager. Senior management needs to recognize the importance of continual improvement and growth in the capabilities of their program managers in order to add increased value to their organization, and invest in career development of their program managers.

Enabler 8: Rewards and Recognition

Human resource policies are needed to support team participation, including a reward and recognition system that is balanced and equitable. This entails a comprehensive approach that focuses on retaining key talent that goes beyond just monetary incentives to also include personal growth and job satisfaction.[10]

Enabler 9: Face-to-Face Investment

Investment must be made in face-to-face meetings that involve critical participants on a program. This will contribute to building trust among team members and form the foundation for ongoing working relationships, improved synergy, and positive team chemistry.

One executive who has experience implementing program management in several organizations is David Churchill, former vice president and general manager of the Network and Digital Solutions Business Unit of Agilent Technologies. Churchill described effective senior management philosophy in support of program management this way:

Organizations that recognize the strategic value of program management will do the following: treat program management as a critical talent and skill set and establish it as a functional discipline like engineering and marketing; elevate the program management function in stature and place it at the senior level in order to provide program managers the necessary level of influence across an organization; and empower program managers as leaders within the organization with sufficient authority to implement and achieve the intended business objectives.

Program Management in Practice

Surviving the Program from Hell

Contributed by Eddie R. Williams

Have you ever managed, or had turned over to you to manage, a *Program from Hell?* You know the kind: highly complex, with diverse stakeholders possessing egos and personal agendas, highly visible to senior executives, involving a huge investment of organizational resources and equally huge expectations, and usually way behind schedule, over budget, and those responsible are acting more like a loosely associated network of people rather than a high-performance team.

Often, the most complex programs are led by the most experienced and seasoned program managers. Sometimes, however, a program manager is thrust into a position to turn around a Program from Hell regardless of experience. Carl Hampton found himself in exactly that position with one exception. He wasn't asked to turn the program around, he was asked to terminate it. However, what was expected to be a sure failure and lost opportunity actually turned into a personal and organizational success. The following is Carl's story.

The Call to Action

Carl was a successful program manager. Over the last couple decades, he had the opportunity to manage several programs and projects, some more challenging than others. One particularly stands out and is now referred to as the Program from Hell.

Carl worked for Healthteq Corporation, a medium-sized software and products firm specializing in health and human services products and had just wrapped up the successful completion of a major program when he was called into the home office on Friday morning to meet with the company's senior management team. During the meeting, Carl was presented the details of a program that was in trouble and was expected to fail. He was informed by management that there were initial overruns, communication issues, problems completing requirements definition, and uncontrolled changes introduced into the system. The result was summarized by one of the executives: "We have an unacceptable number of issues and defects, to the point we can't win with this solution."

It was also shared with him that there were problems between the client and certain program personnel. He was told he would be replacing the current program manager, William Nicks, who would be reassigned.

Carl was told that he was selected for this assignment because of his program management capabilities and more, specifically, his ability to manage relationships and his conflict management, negotiation, and communication skills. Carl had gained recognition for these skills because of his leadership and management on previous programs and projects. The management team acknowledged Carl's abilities, and indicated that this program required someone who was a "business and technical program manager; someone comfortable working with strong personalities on the business side and capable of leading a diverse team of project managers, technical leads, and managers."

The position change would be announced in a week, which gave Carl time to work with the current program manager regarding all matters pertaining to the transition. At the close of the meeting, the director of technology investments said, "Your primary goal here, Carl, is to gracefully terminate this program."

Carl had heard rumors that this program was in trouble. He did not know, however, that he would be asked to terminate the program and he now had concerns about doing so. He was having a great deal of success in his career, and didn't have any failed projects or programs as part of his resume. He wanted to keep it that way. However, being in the central office with the executive leaders in front of him, he said he would get on board and take a good look at the program. Before making plans for the program's termination, he would report back to the executive team in a week with a report on his findings and recommendations for next steps.

Taking a Look at Hades

Carl came on board the program the following Monday morning and met with William Nicks. Carl informed William that he would be conducting a program assessment that would entail interviewing client managers, project team leads, and subject matter experts. The two reviewed the assessment together, which included the following nine points:

1. The contract, proposal, and statement of work
2. Client's business plans and strategies that the program supported
3. Program business case and any charters or briefings that existed
4. Financial expectations and budget reports relative to ROI expectations
5. Program deliverables, project plans, and schedules (plan versus actual)
6. Methodologies in use
7. Initial requirements documentation
8. Program and project status reports
9. Risks, communication, and security plans that existed

During the first week Carl was positioned as the co-program manager and introduced to management, clients, users, other stakeholders, and employees as appropriate. Healthteq management expected William to move on as soon as possible (both Carl and the management were working to help him do so). William worked with Carl to arrange pertinent meetings and to gather and share material needed for the assessment.

Throughout the assessment, all those interviewed were open and honest. However, tension was high among some managers and users on the program displayed some anger. The users, in particular, continually voiced their negative thoughts about the program. After the first two days, Carl could imagine how difficult the past couple of months must have been. He debriefed with William after the second day and said, "It's been a very uncomfortable week so far, and I understand that as the program manager, relationship and conflict management are now my responsibilities."

By the end of the week, after meetings with stakeholders and members of the program team, Carl had identified several issues, problems, and challenges. However, as he documented his findings and prepared his briefing for the executive team the following week, he became convinced that the program could be put back on track and deliver the desired business value rather than take the loss associated with terminating the program.

Carl detailed his thoughts and turn-around plan, arranged a meeting with his program team, and presented his assessment. He clearly articulated his evaluation with recommendations and outlined a series of action plans. His program team acknowledged that the assessment was accurate and that it illustrated his ability to listen to their concerns and ideas. One member of the program said, "We presented similar concerns to William on a number of occasions. I for one am glad you listened to the concerns, validated them, and are helping us with a plan of action."

Carl let the team know that William had accepted a new position. He also said, "With your support, I will carry this assessment and plan to senior management for approval. I will only do so, however, if you are committed to it and will continue to voice your concerns when you see them, and work together to resolve them." His core team agreed. Team buy-in was important because the next steps Carl outlined did not focus on terminating the program. Rather, his next steps focused on turning the program around and delivering the planned business benefits. Carl left the core team meeting to present his findings and turnaround plan to the executive team.

Carl presented his assessment and proposal to the Healthteq management team. They agreed with the assessment and asked him to arrange a meeting with the client as soon as possible. Carl held that meeting and presented his plans to the client and won their support to proceed.

"Right-Tracking" the Program

Carl focused his immediate next steps on conducting coaching sessions with his project managers and team leads. He discussed both problems and successes that he saw with the program. He actively solicited their feedback, discussed his assessment, and made recommendations required for each project. Through this process, the core team members informed Carl that they had previously met with two program team members and had provided warnings regarding their inappropriate communication with the client and users. Their communication skills were not adequate for their current assignments and previous meetings and discussions with the team members did not resolve the issues. The consensus was that the two members be moved to more appropriate assignments outside of the program. Carl helped facilitate these moves.

Although William had ensured that proper kick off meetings had taken place, the clients thought that there also should be more communication and interaction between the program manager and client management team. Carl made the following communication changes:

1. Met with the client on a weekly basis to discuss status, market, and risk. You cannot have too much client communication.

2. Developed a stakeholder plan to include identification, needs assessment, and strategies for managing their concerns with corresponding follow-up during the life of the program.
3. Continued the necessary meetings to coach project managers and team leads.
4. Continued to monitor the program process to ensure it remained aligned with the business and product marketing strategy and planned return on investment—Job 1 is business strategy alignment.
5. Ensured that resource planning and management requirements were met for the program and projects. The program manager has to create a sense of urgency (and rationale) for proper program resourcing.
6. Ensured that knowledge sharing took place and lessons learned were documented and used. Continuous organizational success is created from learning and knowledge sharing within and across programs.

During the reorganization efforts, Carl conducted ongoing communication sessions with his management team and users. To gain their respect and support, Carl ensured that his communication was open, honest, and that he operated with the utmost integrity. He consistently operated in that manner and established weekly communication meetings with management. He met with projects managers, business and technical leads, and other team members to restate the program's vision, goals, objectives, and expectations for each project.

Instead of being completed the following October as planned, the program was successfully closed at the end of June—four months early. This created a large cost saving, but more importantly, the customer was satisfied with the final result and Healthteq realized business value that senior management once thought was a loss.

Keys to Turn-Around Success

Effective communication was the keystone to Carl's turn-around success. However, it was not the only factor. His application of solid program management practices was also important as described below.

Important to any program turn-around effort is putting due diligence into resetting the program on the right trajectory. Carl began with reexamining the business and solutions requirements and gaining mutual agreement with the client on the appropriate requirements going forward.

Additionally, he worked with his core team to establish configuration management, quality assurance, and change management processes for the program. A two-tiered change management process was instituted at both the program level and project team level.

With a new set of requirements in place, Carl created a program work breakdown structure and worked with his team to identify project deliverables, due dates, primary program milestones, and critical deadlines. This information was used to create a new program map, project-level schedules, and revised integrated program plan.

Carl also had to make some resource adjustments. In particular, he had to replace the two team members who were disruptive to both the users and the program team. He worked with the appropriate project managers and functional managers to remove them from the program and replace them with people who possessed the necessary technical and interpersonal skills.

Finally, Carl established a program governance committee to oversee the program and provide direction and guidance through the remainder of the program cycle. For the first nine weeks following Carl's appointment as program manager, he provided weekly program status and performance reports to increase the confidence of the governance committee and senior management team that program turn-around was possible.

Carl successfully turned this program around and delivered on customer expectations and business goals. His primary competencies in doing so were centered on effective communication, collaboration, and negotiation skills. He also demonstrated business savvy and solid program management expertise to diagnose the current situation and to create a plan to "right-track" the program. Importantly, however, Carl had people skills to quickly establish rapport and build trust and credibility with a diverse group of stakeholders. In the end, Carl maintained his successful track record of planning, leading, and delivering program results.

ENDNOTES

1. Crawford, L. "Competence Development." In P.W.G. Morris and J. K. Pinto, *The Wiley Guide to Managing Projects*. Hoboken, N.J.: John Wiley & Sons, 2004.

2. Pink, Daniel. *A Whole New Mind*. New York: Berkeley Publishing Co, 2006.

3. Pinto, Jeffery K. *Power and Politics in Project Management*. Newtown Square, PA: Project Management Institute Publishing, 1998.
4. Martinelli, Russ; T. Rahschulte; and Jim Waddell. *Leading Global Project Teams: The New Leadership Challenge*. Oshawa, Ontario: Multi-Media Publishing, 2010.
5. Goleman, Daniel. *Emotional Intelligence: Why It Can Matter More Than IQ*. New York Simon & Schuster, 2005.
6. Martinelli et al., *Leading Global Project Teams*.
7. Adler, Nancy J. *From Boston to Beijing: Managing with a World View*. Cincinnati: South Western, 2002.
8. Martinelli et al., *Leading Global Project Teams*.
9. Wheelwright, Stephen C., and Kim B. Clark. *Revolutionizing Product Development: Quantum Leaps in Speed, Efficiency, and Quality*. New York: Free Press Publishing, 1992.
10. Martinelli et al., *Leading Global Project Teams*.

Part V

Organizational Considerations

Transitioning to a program-oriented organization to develop new capabilities can be a challenging endeavor. In Chapter 1 we introduced the program management continuum as a way to demonstrate the variations in the implementation of program management within companies today. At the center of the continuum is an important point, which we refer to as the *point of transition*. This is a philosophical decision point where the senior leaders of an organization make a purposeful decision to transition their organization from being primarily project-oriented to being program-oriented. This section focuses on organizational considerations regarding how best to plan and transition to program management.

Specifically in Chapter 12, "Transitioning to Program Management," we outline a process to determine if crossing the point of transition is what an organization really needs. As noted earlier in the book, many organizations are quite successful operating as project-oriented firms. Others, however, recognize that they have a need to cross the point of transition to become program-oriented. Not only is a decision model presented, but importantly, a transition model and best practices in transitioning are detailed in this chapter.

Chapter 13, "The Program Management Office," details the role the PMO plays in establishing program management as a functional discipline within an organization. The description of the program management office, its primary functions, and operational elements are discussed. Additionally, PMO implementation guidelines and a variety of organizational factors worth considering when a management team is evaluating the formation of a program management office are presented.

Chapter 12

Transitioning to Program Management

Organizational transformation is a challenging endeavor. Transitioning to a program-oriented organization is no different. In Chapter 1 we introduced the program management continuum as a way to demonstrate the variations in the use of program management within companies today. At the center of the continuum is an important point that we refer to as the *point of transition*. This is a philosophical decision point where the senior leaders of an organization make a purposeful decision to transition their organization from being primarily project-oriented to being program-oriented.

A transition to program management can yield significant improvement in a firm's ability to achieve their strategic objectives, competitive position, and financial returns. To realize such improvement requires an awareness, willingness, and commitment to change the organization's culture, overcome internal politics, and establish a new mindset regarding roles, responsibilities, and functions. Crossing the transition point to become a program-oriented organization can affect all levels of management and employees. It changes the rules of engagement, decision-making hierarchies, teams and team structures, and requires competencies different than those prevalent in traditional project-oriented organizations.

Our experiences in leading organizations through this transformation have revealed several key factors that must be managed for success, including:

- First and foremost, there must be a compelling reason to change.
- Senior management must drive the transformation and have a clear vision of the end state.

- Changes to the organizational structure may be required, therefore preplanning to do so is necessary.
- People need to understand their new roles and responsibilities and the new power equation.
- New core competencies may be required within the organization.
- Behavioral change on the part of individuals affected has to be modeled, monitored, managed, and incentivized.

This chapter focuses on the management of these factors relative to the process of change. First, we share commonalities of successful organizations that have made the transition from a project-oriented to program-oriented business. Then we provide the critical elements that constitute a successful change transition, some typical challenges that will need to be overcome, and finally the need for continuous improvement.

UNDERSTANDING CHANGE

The goal with transitioning to a program management discipline is to achieve a fully functioning model that yields improved business results and do so with as little disruption and pain as possible throughout the transition. To achieve this goal requires understanding change, risks associated with change, and roadblocks that prevent change. The following are helpful in achieving the goal:

- Before starting the transition, know whether or not you really need to change.
- Recognize that successful change is both evolutionary and revolutionary.
- Create a compelling story for change.
- Establish ongoing communication and feedback pertaining to all vested parties.
- Have senior management actively engaged in ensuring the successful implementation of the change.

The Need for Change

There are a number of key factors that typically drive a senior leadership team to decide that program management is needed within their business. The most common factors we have witnessed in working with organizations are detailed in Chapter 2, and summarized below.

Improving alignment between business strategy and execution output. As an organization begins to grow its business, maintaining alignment between work output and strategic intent often becomes a challenge. Simply stated, the link between execution output and strategic goals begins to weaken, and in some firms may eventually break. For many organizations, program management provides value by serving as the organizational *glue* that can prevent the misalignment between execution and strategy by creating a critical linkage between strategic goals, program objectives, and project deliverables.

Managing increased complexity. The pervasive rise in complexity due to customer demands for more integrated, larger scale and connected solutions has emerged as a primary challenge to the way we have historically operated within our businesses and organizations. For decades, people working to create and deliver complex solutions have used a systems and program-based approach to simplify the levels of complexity they encounter.

Achieving business scalability. As an organization grows, its ability to effectively scale its business processes and consistently deliver positive return on investment can be constrained by a business manager's ability to personally scale. The addition of the program management discipline and skilled program managers removes this constraint as program managers become accountable for delivering the business results for each of their programs.

Improving integration of cross-business functions. Companies continue to be plagued by ineffective collaboration between their business units and divisions caused by parochial behaviors within traditionally siloed functions. The inclusion of program management in a firm's business model can provide value to those organizations by helping them to execute business strategies through the collective efforts of their business functions working to achieve a common vision.

Managing distributed collaboration. The ability to digitize, disaggregate, distribute, produce, and reassemble knowledge work across the globe has created a new business model where highly distributed team collaboration is required. Many companies have come to the realization that they need to adopt a new business model not only to compete, but in many cases to survive. For many, program management provides a management model that specializes in the distribution and reintegration of work.

Table 12.1 Decision table for transitioning to program management.

Factor	Project Management	Program Management
Degree of linkage between execution output and strategic goals	Loosely connected	Tightly connected
Technical complexity	Low	High
Structural complexity	Low	High
Business complexity	Low	High
Level of business scalability	<6 large work efforts	>6 large work efforts
Business function interdependency	Low	High
Multi-site team distribution	Less than 5 sites	More than 5 sites

Now that we understand some important factors involved in the transition decision, we can essentially consider whether to transition to a program-oriented organization. As noted, organizations can find success with either a project- or program-oriented model. Each fits one situation better than another. To decide whether to transition, we devised a simple tool, outlined in Table 12.1, that can be used to guide a discussion and ultimately a decision.

In Table 12.1 we list which of the two approaches may be a better fit based on eight factors. This table can be used as a guide in the assessment, discussion, and decision to transition to a program management model. When using this table, we recommend first tailoring the factors for your specific case, and then determine if each factor favors project or program management by placing an "X" beside your choice.

If the majority of "X's" are on the project management side, do not transition to program management. If, however, the majority of "X's" are on the side of program management, you may have the justification to prepare a case for transition. A word of caution may be of help here. This is a simple but practical tool to make your transition decision. Like any other tool, it is only as good as the information contained in it and the discussion that occurs from it. Therefore, prepare high-quality information for your case and assess the factors completely, critically, and professionally among senior decision makers within your firm.

If the decision is made to transition to program management, we recommend creating a business case for such a transition as outlined in Chapter 6. Use the business factors and operational factors to guide the business case and discussion regarding the transition. Further, don't forget about the need to plan for and overcome communication challenges (build a compelling story), active management support (ensure senior management engagement), and mitigate possible in-fighting due to politics, defending turf, and power tripping.

Transition Is Both Evolutionary and Revolutionary

To obtain the broad-based benefits from applying program management to the business model of the firm, several fundamental changes are required. As noted earlier in this chapter, to achieve a successful transition to program management requires an awareness, a willingness, and a commitment to change the organization's culture, overcome internal politics, and establish a new mindset regarding roles, responsibilities, and functions. This affects all levels of management as it changes the rules of engagement, decision-making hierarchies, and team structures, and requires competencies different than project-oriented organizations. To achieve such change requires time, which is the reason we use the word "transition" throughout this chapter. Transition is meant to imply changes are being realized over time—*evolutionary*. On average, we have consistently witnessed the transition taking three to four years. This is from transition decision to fully functional program management operations.

The successful transition to a program-oriented organization requires the senior management team of the firm to consciously decide that they will make the fundamental structural and cultural changes necessary to move the organization to program management and lead the implementation of changes. This decision is *revolutionary*, as are some of the fundamental changes that are required, such as changes to roles and responsibilities, changes to decision-making empowerment, changes to the organizational structure, and changes to the incentive systems. Therefore, success in transforming an organization to a program management discipline is both *evolutionary* and *revolutionary*.

Create a Compelling Story for Change

Transition success begins with creating and communicating a compelling story for change.[1] A compelling story is necessary because, prior to any change, people must first understand *why* the change is needed.

Beyond just understanding why the change is needed, importantly the employees must agree with the change before actually changing. To accomplish this, you need a compelling story.

There are three points to remember when crafting a compelling story. First, what motivates you may not motivate everyone. Some people may be motivated by a compelling story about how to positively impact society. Others may be focused on maximizing customer satisfaction. Others may be interested in achieving the highest level of financial return. Others yet may be focused on self—what's in it for me. A compelling story will focus on all (or at least multiple) areas of motivation—self, organization, customers, and society.

The second point to remember in creating a compelling story for change is that the story is best written by the employees. Often, position-level leadership (the CEO, for example) writes and conveys the story. The highest probability for sustainable change occurs when employees can write their own story. In this case, the leaders' role is to clarify the situation (problem or opportunity) from which others set the change in motion relative to their story—what we need to do, what we can do, what we will do—and then the leaders establish commitment, responsibility, and accountability to enact the change. This establishes a culture of employees being involved with the change rather than having the change imposed upon them.

The third point to remember in creating a compelling story for change is relative to the problem or opportunity. Sometimes senior leaders get in a pattern of promoting change because of a problem—not enough units sold, not enough market share, not enough profit margin, and the like. Other times, the pattern may be about opportunity—chasing best practices regarding one thing or another. This can be tiring for individuals and organizations at large; the results could be change fatigue in which case creating a compelling story and establishing a sense of urgency lose meaning due to desensitization to the need for change. To be effective, the compelling story needs to balance problem and opportunity.

Senior Management Sponsorship

Firms that have been most successful in transitioning to a program management model have had strong sponsorship from the senior management team of the organization. Implementation of the model works best when it is managed from a top-down approach, rather than a bottom-up approach.

When senior management buys into the approach and drives the implementation, the rest of the organization begins to fall in line. It is difficult to derive the necessary and significant changes in an organization when it is attempted through middle management, because barriers within the functional organizations quite often become too difficult to overcome. Coupling this point of sponsorship with the need for a compelling story, as noted above, senior leaders can push for change (from the top), but are most successful when engaging staff in making sense of the change and detailing the story for change.

One senior executive told us how his first attempt at implementing program management within an organization was a complete failure. As he told us, "The transition failed because it didn't have executive support and, worse yet, the executives weren't even open to learning about the value of program management." This was an organization that was too hard to move without top management supporting and driving the transition and necessary change in culture. According to the senior executive, their philosophy was, "If a person was not a technologist, he or she wouldn't be able to lead a development team." This is a common problem encountered in technology or engineering companies, making the transition to program management much more difficult than in other types of organizations.

Success most often occurs when the senior sponsor is an influential change agent with a compelling story for change. This means someone with sufficient power, influence, and respect in the organization; someone who can ensure that the organizational structure, behavior, and culture move in a new direction. The change agent must have a clear vision of what the change is and why it is important for the success of the organization. This value proposition should be sufficient for the senior executive to champion the change.

EXECUTING THE PROGRAM MANAGEMENT TRANSITION

Before transitioning to program management, a firm foundation has to be established from which to build upon. This foundation consists of a strategically focused management philosophy and strong senior management engagement.

Senior management's philosophy as to whether they choose to view project work strategically and linked to the success of the business, or rather to view the efforts as tactical and operational in nature, is critically important for a successful transition to program management. For a program manager and his or her program team to perform at their highest potential, senior management of the firm must make it apparent that they believe the team's efforts are directly tied to the success of the business and are integral to the strategic success of the enterprise. With this view in place, it elevates the importance and resulting influence of the program managers and enables them to coalesce action to get things done.

Additionally, senior management should position program management as a critical business function within the organization. This will enable transition success. Senior management must be careful to preserve the functional neutrality of the program management function. Firms that view program management as a true functional discipline that is equal in influence to other functions of the organization experience increased program management talent and capability, ensure the long-term viability of the discipline, and potentially develop future business leaders for the enterprise.

The phrase "actions speak louder than words" is common. In a world where most executives espouse values such as integrity, honesty, respect, quality, stakeholder satisfaction, and the like, differentiation cannot be found in the words, but rather in actions. The same holds true when leading change. It is not enough when senior leaders simply agree to a change effort, nor is it enough to simply fund the effort. One of the things that separates industry leaders from followers is a senior management team that is actively engaged in planning, leading, and sustaining change.

Whether you are an executive or not, walking your talk is an individual skill that can create corporate identity and distinction. Importantly, as James Kouzes and Barry Posner noted, when values such as integrity are agreed upon and reinforced throughout an enterprise, productivity increases, corporate loyalty is strengthened, teamwork is embraced, and change becomes easier.[2] Performance—success or failure—is always the product of the actions from people, not the words in a report or on a poster in the foyer at corporate headquarters.

Transitioning Methodologies

Once a strong program management foundation is established, transition to a program-oriented organization can begin. Many day-to-day tasks have to be managed during the transition, but several structural and cultural elements also need to be in focus for a successful transition.

Many methods to explain transformation have been published. Some are highly academic, others are highly technical, some are more useable than others, and all of them have shortcomings. The challenge is that many tools and models for change explain *why* change is needed, yet are rather superficial when it comes to explaining *how* best to transition an organization.

Explaining how to transition is difficult because of the uniqueness of organizations, their strategies, their culture, their situation. It is for this reason—uniqueness—that we recommend having a qualified organizational development expert within the business or an experienced organizational change consultant to lead the transition process and team. Experienced individuals in this field have the requisite skills to navigate over or around the barriers that will be encountered and they represent a neutral position within the company, both of which are necessary to properly plan, lead, and sustain organizational change. Having said that, it is valuable to become familiar with popular change models for reference purposes and for a starting place for planning your transition. Although dozens of models exist, some popular models include Marvin Weisbord's six-box model,[3] William Bridges' three-stage model,[4] and John Kotter's eight-step process.[5] These models and others were studied by Sergio Fernandez and Hal Rainey in a meta-analysis of organizational change models, processes, and practices. They concluded that there are eight common factors among organizational change models.[6] These factors include a common understanding of the needs, a plan, internal support, means to overcome resistance, top management support and commitment, external support, resources, institutionalization of the change, and a desire to accomplish even more comprehensive change.

Our experience is that while these models exist, most attempts to plan, lead, and sustain organizational change fail. Change transition success in reality has little to do with the change model implemented, and much to do about changing behavior and overcoming political resistance. While these models can be used to provide a framework for what needs to be done during an organizational transition, they fall short on advice on how to do it. Based on our experience working with organizations involved in program management transition, we have been able to extract a common and repeatable process used by these organizations. This process is presented in Table 12.2.

What is a bit remarkable about this process is that it has been applied by a wide variety of companies in a number of business settings: a customer

Table 12.2   Best practices for program management transition.

Transition Step	Required Action
1. Point of transition decision	Senior leaders realize a systemic problem is plaguing them (misalignment between execution and strategy, growing complexity, business scalability challenge) and make the decision to embark on a program management transition.
2. Assign a transition owner	A senior leader is appointed as transition sponsor and assigned responsibility for assessing the need for transition and to lead the transition effort.
3. Establish a broad knowledge base	Recognizing program management needs to be part of the business model and company mindset, best practice organizations invest in program management fundamentals training for their personnel to establish a knowledge base (beginning with the senior management team).
4. Modify roles and responsibilities	Each organization is unique with varying roles, responsibilities, policies, procedures, and practices, not to mention culture. Transitioning to program management will require adjustment of *many* of the roles and responsibilities previously established for job functions.
5. Create a center of excellence	Many organizations believe it is critical to create a center of excellence, which is important for gaining a foothold in the midst of change and establishing a means to sustain the change. The center of excellence is responsible for adopting and establishing a set of critical program management processes, tools, and metrics. One method for accomplishing this is through the use of a program management office (Chapter 13).
6. Execute pilot programs	Program management is often implemented in waves. A roadmap for the program management transition should be established that clarifies which projects should be stopped, which will continue uninterrupted, which will be intercepted, and how new programs will be established. The piloting of the conversion of projects to programs is critical to establishing and adopting the new processes and standards (see "Pilot, Learn, Improve"). It is important to learn what works and what meets resistance before fully committing the organization to a broad-based transition.

(Continued)

Table 12.2   (*Continued*)

Transition Step	Required Action
7. Institute organizational structure changes	Most organizations find a need to make adjustments to the organizational structure during the transition process. This is driven by the need to elevate the program managers and program office to a level consistent with other functions to establish authority and decision empowerment.
8. Reinforce the correct behavior	As the transition continues past the piloting stages, rewards and incentives are modified to engrain desired behavior change as part of the new way of doing business.
9. Set expectations for continuous improvement	The center of excellence established to provide early transition direction is rechartered to develop a roadmap for continuous improvement and to drive the implementation.

resource management software company implementing program management to provide alignment of their agile software processes to the business goals of the company, an aerospace company reestablishing program management as their basis for developing their new products after experiencing a significant number of execution failures, a start-up services company using program management to provide structure to scale its business and provide repeatable results, and a nonprofit organization implementing program management in order to manage the growing complexity of its projects.

Pilot, Learn, Improve

Individuals within an organization are only able to comprehend and act upon a finite amount of change to the way in which they perform their jobs. Therefore, transition to a program-oriented organization must begin slowly with a fairly narrow scope to build confidence, credibility, and momentum. It is recommended that pilot programs be utilized initially to gauge the amount of change that can be sustained within the organization and to capture early learnings that will be valuable going forward. Broad-based communication up, down, and throughout the organization is critical during this time frame to ensure that progress, changes, and issues are well understood. Rumors and misinformation may be frequent and numerous, which will require active management by the team.

By taking an iterative process to implementation, it allows the transition team, management, and the broader organization to do the following:

- Start slowly
- Learn by doing
- Build confidence and credibility
- Achieve early successes and demonstrate tangible results
- Seek feedback and involvement of personnel across the organization on a regular basis
- Adjust the plan as needed when things change
- Create new benchmarks and raise the bar for driving continual improvement
- Ensure the internalization of new values and behaviors

Designing the Program Management Model

We mentioned at the start of this chapter that when organizations make the decision to cross the point of transition on the program management continuum (Chapter 1), they make the decision not only to become a program-oriented organization, but importantly they make the decision to transform their organization. Therefore, when planning for program management transition, aspects of structure, system, and culture will be at the forefront. Each is connected and affects the other relative to program management design. This is illustrated in Figure 12.1.

Program Management Structure

Utilizing program management to create new capabilities for an organization involves the combined efforts of many personnel across an organization who work together to perform the tremendous number of activities and tasks necessary. This combined effort requires effective communication, coordination, ability to resolve issues and barriers, and effective decision making both within the team and by senior management. An organizational structure that supports and facilitates this collaborative approach is necessary.

Traditional organizational structures range from a purely functional structure to a purely project-oriented structure. Both extremes create some significant limitations from the perspective of program management. The purely project-oriented approach minimizes the critical importance played by functional managers in providing for the long-term viability of the functional capabilities by staying current on the latest technology

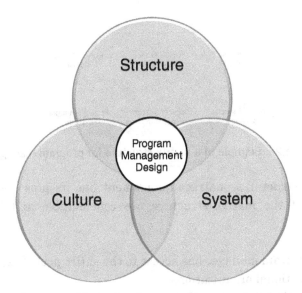

Figure 12.1   Elements of program management design.

and maintaining the highest-skilled and best-trained resources to support the enterprise. A purely functional organization minimizes the importance of cross-discipline knowledge and a systematic view needed for programs. Functional organizations tend to limit cross-discipline thinking, with authority resting only with the functional managers. As Christopher Meyer, author of *Fast Cycle Time*, stated, "By far, the most serious problem with the functional structure is that it serves its members better than its customers."[7]

The most effective organizational structures for the program management model is a compromise between the two extremes—the matrix structure and the program structure. With both approaches, formal responsibility and authority for the program resides with the program manager.

The Matrix Structure   Figure 12.2 illustrates one form of the matrix structure that is used in various companies, but there are, of course, many variations of the matrix structure that can be implemented.

In a matrix structure, program team members continue to report directly into their functional organization (depicted as solid lines) and are loaned to the program manager (depicted as dotted lines). The program manager is responsible for integrating the cross-discipline contributions. The functional managers are responsible for overseeing the core capabilities, processes, and tools within the function.[8]

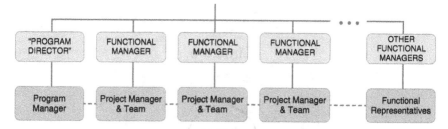

Figure 12.2 Example of a matrix structure for program management.

Like all structures, matrix management has its pros and cons. The strengths of the matrix structure are considerable and include the following:[9]

- The program manager has access to the entire pool of experts within the functional organizations.
- Duplication of functional departments is eliminated, with a single organization for each function.
- Resources can be shared intelligently across programs to ensure effective use of critical and scarce skills.
- Integration of the functional disciplines is enhanced, which serves to reduce power struggles and improve collaborative teamwork.

The Program Organization Structure The second organizational structure that we will discuss is the program organization structure shown in Figure 12.3. This structure comes in many forms and is found in industries such as aerospace, defense, and automotive.

In the program organization structure, the project teams report directly to the program manager. Whether functional managers report directly to

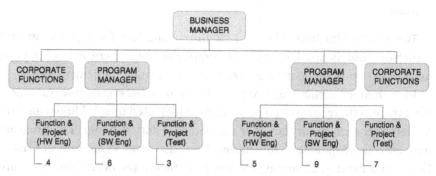

Figure 12.3   Example of the program organization structure.

the program manager is specific to the organization. All resources reporting to the program manager are dedicated to and work full-time on their programs. Generally, cross-program functions such as marketing, finance, and human resources support all programs, and therefore are not dedicated to a single program.

Strengths of the program organization structure include the following:

- The program manager is in direct control of all resources on the program.
- Resources are completely focused on a single program, which leads to improved productivity and cycle time.
- A high degree of cross-project integration occurs.
- A high level of program identification and team cohesion emerges.

An important element in both organizational structures is that program management is a stand-alone function. This is important to the program management model because the positive impact of program management is considerably diluted when it is contained within an existing functional organization.

Program Management Systems

When transitioning to program management, multiple systems need to be considered and modified as needed. First, program management needs to be integrated into the business and operational systems of the company as described in the integrated management system (Chapter 3). Most important is that there is systematic alignment among business strategy, the portfolio management process, and program management practices. One of the biggest transitions for any company making the transition to program management will be a restructuring of the portfolio from a set of independent projects to a set of programs with constituent projects. Additionally, relative to the operational system, alignment between business benefits, program objectives, and project performance indicators has to be established and managed for each program.

Program management transition will also bring a change in core practices, processes, tools, and metrics as described in earlier chapters. In some cases the change will be primarily an up-level or abstraction from a project level to a program level, such as in the case of risk management. In other cases new practices, processes, tools, and metrics will need to be developed and instituted as in the case of business benefits management. It is important to note that we have seen companies consistently struggle with transitioning their metrics system. Generally, firms put more emphasis on the transition of practices, processes, and tools, but seem to want to hold on to their existing set of measures and metrics. As practices change, so too must the system of metrics supporting the practices.

As indicated earlier, a company's system of roles and responsibilities will also most likely change as a result of a transition to program management. This is due to the fact that the role of the program manager is being inserted into the organization, or it is being changed to encompass aspects of other historically established roles. This realignment of roles and responsibilities can be tricky and needs to be carefully managed (see "Roles for Effective Program Management").

Roles for Effective Program Management

Well-defined roles for the management team involved with a program are important to clarify program governance and decision making, lines of communication, and program responsibilities.

Senior Management: The role of senior management is to provide strategic guidance for the program and to establish and maintain the processes and procedures by which programs are managed within their organization. Senior management owns the portfolio of programs for the organization; selects and empowers the program manager to form a team and organize a program; assists the program team in setting the key objectives, targets, and constraints for the program; makes key approval and direction setting decisions; and provides the necessary leadership and guidance on issues and barriers outside the control of the program team.

Program Manager: As detailed in Chapter 10, the program manager has three primary roles. First, the program manager must manage that portion of the business represented by his or her program as a proxy for the business manager. Second, the program manager manages the cross-project interdependencies to ensure that an integrated solution is created and delivered. Third, the program manager must lead the team of people empowered to him or her to achieve the program vision and business benefits anticipated.

Project Manager: The project manager serves on the IPT core team as the leader of the function he or she represents and manages the individual contributors on the project team. The project manager drives the creation of the project plan and integrates that plan into the overall program plan. He or she then drives the implementation of the project plan, manages the work commitments within the project, and ensures completion and closure of all project deliverables. The project manager works closely with the program manager and other IPT core team members to seamlessly integrate all functional elements of the final solution. The project manager is also the key conduit for communication on the program between the program team and the functional organization he or she represents.

Functional Manager: The functional manager is responsible for the hiring, training, and development of the functional personnel. He or she is also responsible for the maintenance of the skills, capabilities, best practices, and tools to sustain the long-term functional expertise. The functional manager makes the resource and work commitments for functional activities in support of the program and assigns a project manager and functional specialists to represent the function on the program.

Program Management Culture

The third element of the program management design (Figure 12.1), culture, consists of organizational culture, power, and beliefs. No matter what approach is taken to handle the culture element of program management, a conscious effort should be invested to purposefully design it, and not just let it evolve.

The program culture should be designed to express a set of clearly articulated, performance-oriented values and on-the-job behaviors that are built into program practices. For example, one such value is "being proactive," in which team members practice a behavior of periodic progress reporting that includes predicting the business results, schedule, and budget at completion of their tasks. The intention is that program team members have a sense of identity with the cultural values and accept the need to invest both materially and emotionally in their program. This should make them more engaged, committed, enthusiastic, and willing to support one another in accomplishing the program goals.

Additionally, program managers should operate under a set of expected values and behaviors that are consistent with program management culture and organizational culture. Typical values and behaviors include the following:

Willingness to lead: Program managers need to take a leadership role to ensure that the viewpoints of all functions and disciplines involved with a program are considered and that work is focused toward achievement of the program vision. Willingness to lead also means being capable of quickly rewarding the team when it is doing well and also possessing the toughness required to manage through problems and conflicts.

Take ownership: Program managers need to display total ownership for the management and outcome of the program. They need to be ready to take total responsibility for their actions and the actions of their team. Additionally, program managers should champion the program within and outside of the organization.

Empower the team: Program managers must be willing to share the empowerment granted to them with the project managers on the program. Likewise, project managers must share their empowerment with their respective team members. Full team empowerment is a powerful tool for quick and effective decision making, where people closest to the issue are able to evaluate the situation and decide proper course of action.

Exhibit balance and fairness: Program managers need to exhibit balance and fairness between cross-discipline representatives on the program team. This means being able to set aside any historical bias toward or against a particular person or function that they may have had in the past. Exhibiting fairness also means setting the expectation that all others on the program team behave in the same manner. Failure to do so can foster unhealthy team dysfunction.

Be persistent: Program management is hard work. During the life of a program, the number and wide variety of problems and barriers that are encountered are enormous. The job is not for the faint of heart. It can take a tremendous amount of drive and persistence to achieve program success.

Extreme care must be taken to align program management culture with the overall company culture. As Roger Lundberg, former PMO director of the Jeep division at DaimlerChrysler explained:

> There is no single right organizational solution. Each company implementing program management will have to align it with their company's culture.

If the program management design does not align with current company culture, there will be human resistance to the transition. Resistance may not only result in cultural difficulties, but to put it honestly, it may terminate the transition to program management if the resistance and cultural change are not properly managed.

OVERCOMING CHALLENGES

When transitioning to a program-oriented organization, to expect there will be no challenges can be foolish, as challenges most likely will be many. From our experience and research, the most common challenges one can expect are resistance to change, tension between program management and functional management, and a misunderstanding on the part of program managers about their boundaries of authority.

Resistance to change is not unique to the transition to a program-oriented organization, as resistance arises any time a significant change to an organization occurs. One can expect resistance from both middle management and some individual contributors. A key contributor to this resistance is that any change is a change from an employee's current status quo, and most times change from status quo breeds resistance due to fear of the unknown. A company may have operated under a business model that generated good to excellent results for years; therefore, people within the organization build their comfort zones around the existing business model. However, if the current business model has run its course of effectiveness and is no longer enabling a firm to keep pace with industry changes, the business and the people it employs must either adapt to the changing environment or prepare for a potentially painful decline of their enterprise. Resistance is likely to occur more frequently and severely in the early days of the transition.

A shift in the balance of power from functional or department managers to program managers commonly leads to tension between the two groups. Many organizations transition from a functional structure in which the functional managers possess resource, budget, and decision-making power to the program management model in which budget and decision-making power shifts to the program manager. Any shift in organizational influence can cause people to react unpredictably. Christopher Meyer called this "the golden rule of organizations"—those who have the gold (in this case, the power) make the rules.[10] For the program-oriented organization to work successfully, this power must be appropriately redistributed between the functional managers and the program managers by the senior leadership team of the organization. The key distinguishing point is that the program manager needs to be formally empowered by senior management to be responsible for all operational and financial aspects of their program.

TRANSITIONING TO A PROGRAM MANAGEMENT OFFICE (PMO)

Many program-oriented organizations may choose to support their program management discipline by implementing a program management office (PMO). This, of course, is a major organizational change for the firm and must be appropriately planned and managed as part of the program management transition process. Businesses that have had the most success when implementing a PMO have consistently followed a comprehensive methodology as discussed earlier in this chapter.

When implementing a PMO within a company, much thought must go into how the PMO will be organized as it may have implications on current roles, responsibilities, and decision making within the organization. How the organizational elements are designed will have a large impact on the overall success or failure of the PMO as an organizational entity. There are a number of important factors that need to be considered when transitioning to a PMO including the following:

Timeline: Implementing a PMO will take time. It is not recommended that an organization attempt to establish a PMO without a transition plan in place. A logical evolutionary plan is suggested for adopting the elements of the PMO as they make sense and the organization's skill level and experience can successfully absorb the new entity.[11]

Centralization Philosophy: What is the firm's management philosophy regarding centralization versus decentralization of various functions and activities? Most organizations would agree on the benefit of centralizing processes, policies, tools, methods, and training. However, controversy may arise when business unit managers or functional managers are reluctant to relinquish the planning and control of programs that are directly tied to the manager's future success.

Reporting: The question as to whether all program managers should report to one person can also be controversial. This may need to be decided and resolved at the executive level. As stated earlier in the chapter, both the matrix and program organization structures work well with program managers who report directly or by dotted-line to the PMO director, while being accountable to the senior management of their business unit for the program results.

Gaining Leadership: As we discuss in Chapter 13, the selection for the PMO manager is critical. He or she will be accountable for achieving the desired business results, therefore clarity of role needs to be established as well as exercising patience to recruit and hire the *right* person.

Continuously Improve: The PMO must be committed to continual improvement and maturing the organization in order to remain a viable and valuable function within the organization. Long-term improvement involves capturing and innovating new best practices, evaluating and implementing more effective tools when they become available, and establishing and evolving a central knowledge base for program management. Also, as corporations continue to become more globally dispersed in their activities, the program management office must become increasingly effective in establishing and improving the linkages to the dispersed sites and closing any communication, collaboration, and operational gaps that may exist.

THE CONTINUOUS IMPROVEMENT JOURNEY

Program management transition is a journey not an event. Without continuous improvement in program management practices, processes, tools, metrics, and behaviors, the program management discipline will gradually deteriorate.[12] Such was the case in many of the industries that were historically strong in program management—automotive, defense, and aerospace. The disinvestment in program management and the resulting atrophy in their capabilities have resulted in mounting execution failures and missed business results in many companies. At present we are witnessing a renewed focus on rebuilding program management capabilities within these industries. Avoiding such a predicament and instead developing a continuously improved program management discipline can be achieved through a number of steps.

Forming a continuous improvement team consisting of a small number of forward-thinking individuals within the organization is necessary to gain momentum and scalability. Much of the team must come from the program management ranks. Having a program management office, even a single-person office, can provide great benefit in leading the continuous improvement team. The team must be capable of looking beyond the transition problems of *today* and create a roadmap for a more effective program management discipline *tomorrow*. The continuous improvement team must be able to focus on the critical systemic improvements needed.

Ideally, there should be a continuous stream of suggestions and ideas for improvement. One tried and true mechanism for identifying systemic improvement needs is the program retrospective.[13] Establishing a consistent practice of holding retrospective reviews will lead to the identification of many of the improvements needed. Best practice companies in this area not only perform retrospectives at the end of a program, but throughout the program cycle. Other mechanisms for collecting improvement needs include direct program team feedback through surveys, focus groups, and one-on-one discussions. Additionally, sharing best practices with other companies has proven helpful for many program-oriented organizations.

Execution of the improvements should be approved by the program governance body, and then implemented within the organization and on targeted programs in a methodical manner. Improvements should be implemented as quickly as possible to reap benefits, but care must be taken to prevent more change than the organization is able to absorb at any one time.

David Churchill, former VP and General Manager at Agilent Technologies, described what he believes to be the critical factors in successfully transitioning an organization to program management:

> Organizations that succeed in transitioning to program management will do the following: Treat program management as a critical talent and skill set and establish it as a functional discipline like engineering and marketing; elevate the program management function in stature and place it at the senior level in order to provide program managers the necessary level of influence across and organization; and empower program managers as leaders within the organization with sufficient authority to implement and achieve the intended business objectives.

ENDNOTES

1. Aiken, Carolyn, and Scott Keller. "The Irrational Side of Change Management." *McKinsey Quarterly* 2 (2009): 100–109.
2. Kouzes, James M., and Barry Z. Posner. *The Leadership Challenge*. San Francisco: Jossey-Bass, 2002.
3. Weisbord, Marvin R. *Organizational Diagnosis: A Workbook of Theory and Practice*. Reading, MA: Perseus Books, 1978.
4. Bridges, William. *Managing Transitions: Making the Most of Change*. Reading, MA: Addison-Wesley, 1991.

5. Kotter, John P. *Leading Change*. Boston: Harvard Business School Press, 1996.

6. Fernandez, Sergio, and Hal G. Rainey. "Managing Successful Organizational Change in the Public Sector." *Public Administration Review* 66, No. 2 (2006): 168–176.

7. Meyers, Chris. *Fast Cycle Time: How to Align Purpose, Strategy, and Structure for Speed*. New York: Free Press Publishers, 1993.

8. Gary, Clifford F., and Erik W. Larson. *Project Management: The Managerial Process*. New York: McGraw-Hill Publishing, 2005.

9. Larson, Eric W., and David H. Gobeli. "Organizing for Product Development Projects." *Journal of Product Innovation Management* 5, No. 3 (1998): 180–190.

10. Meyers, Chris. *Fast Cycle Time: How to Align Purpose, Strategy, and Structure for Speed*. New York: Free Press Publishers, 1993.

11. Archibald, Russell, D. *Managing High Technology Programs and Projects*, Hoboken, N.J.: John Wiley & Sons, 2003.

12. Juran, Joseph M. "Managing for World-Class Quality." *PM Network* 6 (1992): 5–8.

13. Lavell, D., and Russ Martinelli. "Program and Project Retrospectives: An Introduction." *PM World Today* 10, No. 1 (January 2008).

Chapter 13

The Program Management Office

The formation and use of a program management office (PMO) is a natural progression as a program-oriented organization continues to grow and mature in abilities and practices. The PMO addresses three of the most common problems that arise as the use of program management increases within an organization. First is the realization of the need for applying uniform program governance and other standards for all programs within a business. Second, the PMO drives consistency and commonality in the use of program management practices, process, tools, and metrics to drive more predictable and repeatable results. Finally is the need for a single focal point for the program management responsibility and accountability within the organization. The PMO leader becomes the individual accountable for overseeing that programs achieve the organization's business results and benefits.

The PMO provides leadership and infrastructure for managing and controlling multiple programs. It represents a compilation of program management infrastructure, tools, and best practices that have been melded together to improve business results and drive continual gains in a firm's program management effectiveness.

Roger Lundberg, former PMO director of the Jeep division at DaimlerChrysler, had the following to say about implementation of program management:

> Successful implementation of program management within an organization requires a certain amount of infrastructure. Key elements of that infrastructure include a good portfolio management process, a well-defined and executed development process, a good requirements system, and good processes such as resource management, change

management, decision-making, budgeting, scheduling, and financial target-setting.

An effective PMO provides the right level of program management infrastructure and aligns it with the overall company culture and structure.

This chapter focuses on the role the PMO plays in establishing program management as a functional discipline within an organization. The description of the program management office, its primary functions, and operational elements are discussed. Additionally, PMO implementation guidelines and a variety of organizational factors worth considering when a management team is evaluating the formation of a program management office are presented.

CHARACTERIZING AND DEFINING THE PMO

So what is a *program management office* and how does it differ from a project management office?

The PMO is a centralized unit, division, or body of professionals within an organization that is responsible for instilling structured leadership, methodology, and practices for programs to make the best use of a company's time, money, and human resources (see "PMO: A New Idea?"). One of the key purposes of the PMO is to ensure that a firm's programs are consistently adhering to the established corporate governance policies, processes, and procedures. The objective of the PMO is to promote and drive consistent and repeatable program management practices within an organization. More specifically, it is a central function utilized for simultaneous management of multiple programs.

PMO: A New Idea?

The first program office was formed and organized in 1957, then called the Special Project Office (SPO), within the United States Department of the Navy. The SPO was established to manage the development of an underwater ballistic missile launch system. Indeed, the structure of the missile launch system program mirrors the program management structures utilized today—a series of interrelated projects (launcher, missile, guidance, installation, navigation, operations, and test) collectively and coherently managed as a program. So, to answer the question from our title—the PMO is *not* a new idea.

However, there is a fundamental difference between that PMO from 1957 and today's offices. The office from 1957, and the wave of offices that followed, had a common attribute of supporting one program only, and certainly those programs were large.

The new wave of PMOs supports multiple programs. This is, of course, an issue of productivity, and resource and cost efficiency. Additionally, these multiple programs are smaller than the likes of the development of an underwater ballistic missile launch system. So, from this perspective, today's PMO *is* new in that it supports multiple programs.

The program management office differs from the more common project management office in at least two ways.[1] First, the program management office is focused on consistency of methods, tools, and practices for all *programs* within an organization, as well as consistency of methods, tools, and practices for all *projects that make up a program*. Therefore, the program management office encompasses both program and project management practices.

Second, the program management office is focused on both business and operational success, whereas project offices are primarily focused on operational success. The PMO is responsible for defining and measuring program metrics that measure progress relative to business goals and ensuring each program supports and is helping to achieve the strategic goals of the company.

In essence, a PMO establishes a formal management function. It provides a true *departmental* home for the program management discipline within an enterprise. This is similar to other organizational disciplines such as human resources, information systems and technology (IT), marketing, and engineering. Each is a functional discipline, unit, or division with the organization and services the entire enterprise.

To be sure, some organizations have established project offices and program offices within other departments of the organization (usually within engineering, IT, or marketing). The inherent functional bias that exists when program managers report into a functional discipline such as engineering or marketing is eliminated when the PMO is established as a peer organization to the others. The PMO provides an established formal path for escalation of issues, barriers, and unresolved risks to senior management for appropriate assistance in resolution. It also facilitates the sharing of knowledge and learning for the organization's program managers.

Program management offices are used in a wide array of organizations and industries for coordinating various types of programs. It can manifest itself in a variation of implementations ranging from an informal community of practice to a comprehensive, business enterprise-level program management office. The appropriate implementation of the PMO is unique and will be dependent upon the objectives, complexity, and culture of the specific firm. Normally, a robust PMO is more useful and cost-effective for larger firms and is especially beneficial for companies with geographically distributed sites and teams.[2] Smaller, single-site organizations, or those with a limited number of programs may not warrant the need for a comprehensive PMO. In these organizations we commonly see either an informal PMO in the form of a community of practice, or an individual serving as the PMO leader.

In addition to knowing what a PMO *is*, it is equally important to know what a PMO is *not*. It is not a cure-all for poor business strategy and it will not substitute for a lack of portfolio planning. Likewise, it is not a strategic planning function for a business unit or a company. Even though it can be formalized within the organization, its structure and use does not have to be rigid or inflexible. It often is not as effective in a decentralized organization, and it is not a panacea that will solve all the problems that exist within an organization.

PMO OPTIONS

The impetus for the creation of a PMO most often originates out of the need for continual improvement in achievement of business objectives and is often driven by major change within the organization. This could be fueled by such things as a firm's rapid growth, mergers and acquisitions, or re-location of operations to multiple sites, to name a few. Changes such as these can result in a greater number of programs, larger programs, increased complexity due to design and team structure, and co-development of new solutions involving multiple geographical sites. These changes, although strategically important to the firm, may create challenges and barriers to achievement of business goals due to inconsistency in program performance; poor communication among project teams, program managers, and key stakeholders; variability in program measurement techniques; and the lack of central coordination of all programs within an organization.

These challenges become even greater as growth occurs internationally, adding language, cultural, and time zone barriers. If these challenges and barriers are not properly addressed, they may result in a firm's inability

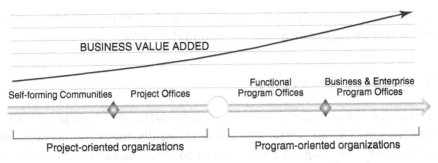

Figure 13.1   Aligning PMO type to organizational need.

to manage and control development efforts, leading to adverse impacts on business results due to missed schedules, program cost overruns, customer dissatisfaction, and competitor gains.

Understanding the type of PMO that is needed is an important aspect of establishing and growing the program management discipline in a firm. Simply stated, the PMO type must align with the current maturity level of a firm's program management capability and have the capability to lead the organization's program managers to a higher level of maturity and performance. This alignment of PMO type to a firm's program management capability is illustrated in Figure 13.1, where we demonstrate the use of the program management continuum to guide the type of program or project management office to implement.

Understanding how program management is implemented within an organization leads to the type of PMO needed. Project-oriented organizations will typically range from seeing no need for a centralized function to guide project governance and standards all the way up to a formalized project management office. These project management offices primarily focus on the operational needs of their projects and programs, and ensure that policies, standards, and tools are being followed. On a more informal basis, project managers in some organizations will group together on their own to form *communities of practice* in order to share best practices, use of tools, and other means to seek continual improvement in the performance of their roles.

A program-orientation means that an organization has philosophically accepted the importance and need for the use of program management as one of the means for achieving their business benefits. In many cases, firms also adopt a PMO approach as a tool for implementing continual improvement to see further gains in alignment of programs to business strategy. These PMOs may range from a functional program office, a business unit-specific program office, or a company-wide enterprise PMO addressing the program needs from the corporate level.

The Project Control Office

The project control office is normally established to provide administrative and tracking support for project teams. Generally, these are independent projects and not integrated nor reliant upon one another for deliverables. In other words, these projects are not part of a broader program.

Work in the project control office is focused primarily on maintaining project procedures, schedule maintenance, earned value tracking, tool usage and support, and project metrics and report generation.

The Functional Program Office

The functional program office is established to support the program managers within a single department, but does not operate as a service or function for an entire business unit or company. Rather, it normally supports a single function such as IT, and is more administrative than strategic in nature. The operations of the functional program office focus on maintenance of program schedules, program and project data tracking, and development of program indicators and reports. The functional program office also maintains a central repository of program and project information pertaining to the particular function in which it operates.

The Business Unit or Enterprise PMO

According to some experts, the establishment and use of an enterprise PMO is a measure of organizational maturity, where program management is recognized as a true function within the company.[3] The key difference between a business unit PMO and an enterprise PMO is span and control. Depending upon the needs of the organization, senior management may choose to focus the PMO on a specific business unit or division within the company, while in other firms there may be a need to focus the PMO at the corporate or enterprise level. Both PMO types, however, are a part of the business engine of the company (Chapter 3).

A business unit or enterprise PMO is the center for program management competencies and practices within a company. It should be established at a level within the organization comparable to other critical functions. Placing the PMO high in the organization hierarchy is critical for two reasons. First, the PMO leader needs to be part of the senior management team of the organization in order to properly align programs to business strategy. Second, the PMO leader must have sufficient political and decision-making influence to manage conflicts and other tensions between various functional disciplines. The business unit or enterprise PMO is normally led by a manager or director who is responsible for all aspects of the program management discipline for the company.

Companies that perform much of their work in a distributed or virtual environment that spans multiple sites and geographies derive a tremendous benefit from the leadership, monitoring, and control offered by a business unit or enterprise PMO. It provides continuity for the portions of a company that are located in geographies far removed from the main offices of the firm that may be experiencing difficulty effectively linking work activities and results with business strategy and staying consistent with requirements established by the home office of the corporation.

An effective business unit or enterprise PMO can have significant influence throughout the company. Figure 13.2 illustrates how it can be used to influence critical aspects of a business' operation such as building a stronger program link to strategy, building program management maturity, forging strong relationships between the program managers and the

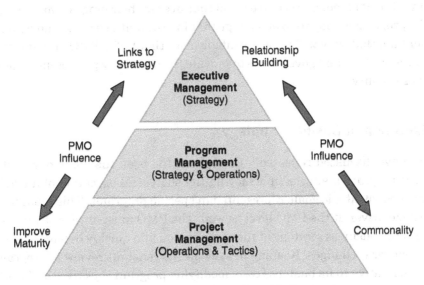

Figure 13.2 PMO organizational influences.

functional and senior managers of the business, and establishing commonality in program management practices, processes, metrics, and tools.

ADDING VALUE TO THE ENTERPRISE

Figure 13.1 illustrates that the amount of influence a project or program office has on business operations is dependent upon the type of office that an organization chooses to implement, with the greatest level of influence and resulting business value coming from an enterprise-level PMO. The following is a description of the value proposition firms have realized by implementing an effective program management office.

Repeatable Business Results

As stated in the PMI *Standard for Program Management* (3rd edition), "establishing an appropriate collaborative relationship between individuals responsible for program governance and program management is critical to the success of programs in delivering the benefits desired by the organizations."[4] In many firms, the program governance accountability becomes one of the key responsibilities of a PMO, and program governance is focused on delivery of business results in a repeatable manner. Business managers need a level of confidence that when they turn their investment of money, resources, and time over to their program managers, the program managers will manage that investment in a predictable manner (and that they will do so repeatedly over time). The PMO provides the infrastructure and governance to provide the necessary predictability and repeatability.

Execution of Business Strategy

We have discussed repeatedly throughout the book the value of a proficient program manager in achieving the successful execution of a firm's business strategic goals represented in specific programs. This must be a critical driver at the PMO level as well. The PMO must always be aligned with the business strategy of the organization and quickly become in sync whenever it changes. Keeping in sync means constant review of program work relative to its business case, modifying program objectives and work efforts if needed, and program termination in some cases. The use of a

PMO can further increase this capability and focus by tightening this linkage between strategy and execution through continual improvement and to carry some of the portfolio management responsibilities for the firm.

Business Risk Mitigation

Informed risk taking is the goal of effective risk management. Significant unmanaged risks can lead to substantial loss or missed opportunities for the organization. Direct involvement by the PMO helps ensure that program managers are fully informed of and engaged in managing the business risks associated with their program, and can take the necessary actions with their team and within their span of control to properly address and manage them.

Management of Complexity

The elements and changing landscape of complexity in today's business environment was discussed in detail in Chapter 4. One of the most significant complexities is managing the geographically dispersed team environment due to the distance, cultural, and communication barriers to name a few. Many companies have made significant gains by applying their PMO capabilities to resolving these challenges. The ability to apply the program management skill and expertise uniformly across all programs, regardless of geographical location and level of complexity, creates a level of synergy that is difficult to replicate by each program team working independently.

Establish Program Management Excellence

Organizations that are program-oriented will many times establish a PMO as *a center of excellence* for the program management function. These firms will seek, either internal or external to the organization, a qualified individual to lead the PMO who possesses substantial program management experience and demonstrable knowledge and understanding of program management methods, processes, and tools. Normally, program managers within the organization will have ongoing accountability to the leader of the PMO for their performance on their programs pertaining to leadership and program management expertise. The PMO provides momentum for continued improvement in program management practices.

KEY RESPONSIBILITIES OF THE PMO

Regardless of its structure, the PMO is the central function for the program management discipline within a business. In its fully developed state, it should contain all elements necessary to ensure that programs are managed effectively and efficiently. If these elements are centrally managed, coordinated, and integrated into a systemized approach, it provides for unified coordination and control while simultaneously enhancing communication across all organizations, functions, physical sites, and time zones. Figure 13.3 illustrates the eight key aspects of an effective PMO.[5]

Program Management Organization

A PMO needs to be established as a true function within an organization and have a well-defined structure with clear roles and responsibilities communicated across the firm. This is especially critical for larger organizations that have numerous program managers. We discussed the

Figure 13.3   Responsibilities of a PMO.

need for empowerment of program management by senior management in Chapter 11, which implies that in establishing a PMO for the firm, senior management is empowering it and supporting it to operate as chartered. This support should be formally communicated throughout the organization and senior managers need to be visible in their support of the PMO and program management processes. One of the important roles of the PMO leader is to develop and administer a career ladder for the growth and advancement of program managers within the firm. The career ladder should include detailed job descriptions outlining responsibilities and experience levels that range from entry-level to senior-level positions. These can be used for hiring, setting performance expectations, promotion, and other professional needs.

Operational Management

Many PMOs provide operational and administrative assistance to the program managers in the organization. This may involve coaching and assistance on utilizing various program methods, processes, and tools as required. The PMO also provides consultation and mentoring for sensitive issues and problems, program start-up and closure activities, as well as a structured process for the escalation of issues and problems. The PMO leader will also be required to interface with senior management, functional management, and other stakeholders within the organization relative to program management and other operational barriers and issues that surface on an ongoing basis as multiple programs are underway.

Governance

Governance, as stated earlier in the chapter, is one of the important responsibilities of the PMO. At the business or enterprise level, governance of programs is absolutely critical. Programs require a more complex governing structure than projects because they involve fundamental business changes and expenditures with significant bottom-line impact. In some instances their outcomes may determine whether the enterprise will survive as a viable business entity.[6] The PMO leader needs to work closely with the designated personnel in the firm responsible for establishing, managing, and administering the corporate governance processes and requirements.

Program Information System

The PMO should be the focal point for program management information and knowledge management within an organization. Centralized storage and organization of program information improves accessibility of the information for the broader organization. Program information includes historical data on program performance to business results, risks, issues and problems encountered, and best practices employed on previous programs. This central repository also allows the opportunity for wide distribution of program retrospective information collected to enhance communication of key lessons learned.

The program knowledge base should include a repository of the common methodologies and tools to be employed on all programs. This includes templates, checklists, and instructional material. Additionally, research, development, and innovation of new methodologies, processes, tools, and technologies are driven by the program management office.

Methods, Processes, and Tools

A primary objective of the PMO is to promote consistent and repeatable program management practices that result in efficient use of business resources. This is accomplished in large part through the development and use of common methodologies and processes. For example, all programs should be managed via a common life cycle methodology as well as consistent scheduling, risk management, change management, and requirements management methods and processes (Chapter 7). These should be common across all sites and geographies to improve coordination and communication to achieve improved predictability of performance on all programs, more effective decision making, economies of scale with respect to infrastructure and tool deployment, and increased flexibility of resources from program to program.

The PMO is responsible for evaluating, selecting, deploying, and many times maintaining the tools that are implemented by the program teams (Chapter 9). This also includes training and mentoring on the effective use of the tools. Like program methodologies and procedures, the PMO looks to drive commonality of tool deployment and use across all programs within an organization. This increases economies of scale to support, use, and maintain the suite of tools.

Metrics and Tracking

The PMO is responsible for developing and maintaining metrics that accurately measure program progress and reflect the achievement of the business goals intended (Chapter 8). This includes trending of the information as needed and publishing the results of the information to the appropriate audiences. The PMO also tracks, collects, and analyzes the data necessary for the development of the metrics through common program tracking tools. New tools are consistently being developed to more adequately monitor and track progress of programs within the simultaneous, multi-program environment. The role of the program management office is to consistently develop the appropriate measures and metrics so the tools provide improved data collection and information dissemination.

Management Reviews and Reporting

The PMO may be responsible for orchestrating and running the formal program reviews for senior management. This includes defining the format of the meetings, the agendas, and the attendees. The PMO provides for an organized and structured approach for management visibility of programs underway in the firm. The structure of management reviews, information to be presented, and who attends the meetings should be negotiated between the PMO leader and other senior managers. The PMO is then responsible for scheduling and leading the reviews and capturing decisions reached and actions needed from the meeting. The focus of these reviews is to determine the status of the programs toward achieving their business objectives.

Competency Development

The program management discipline will only be as effective as the program manager's skills, capabilities, and experience. Skills training and development can and should be made available to all program managers, and be part of the PMO charter. Much of the training material can be maintained on websites and available to all personnel, and can include online modules as well as face-to-face group sessions and forums addressing specific topics. For multi-site firms, training and best practices for managing in a distributed or virtual team environment can be provided.

Training and career development of the program managers in the organization is driven and supported by a program management competency model as discussed in Chapter 11.

The PMO is responsible for creating a learning environment through the alignment of all program managers and their teams under the same leadership, philosophies, and practices. Program managers master their skills as they communicate and coordinate their programs with other program managers in this structure and, therefore, are continually observing and learning from one another.

THE PMO LEADER

It is not too surprising that when one envisions the role of a program manager as difficult, they also view the role of a PMO leader is an order of magnitude more challenging. PMO leadership is a critical role in a program-oriented company. As was stated by Lia Tjahjana, Paul Dwyer, and Mohsin Habib in their book, *Program Management Office Advantage*, "an organization's success in operating a PMO depends primarily on the competencies and performance of its staff and most specifically on the hard and soft skills of the PMO leader."[7] The case study provided in Appendix A ("I AM the PMO") provides a good sense as to how a PMO can be an integral part of a corporation and how establishing this role can be so challenging.

The role of a business unit or enterprise PMO leader can carry a wide array of responsibilities, ranging from management of the program managers and related governance, processes, methods, and tools, to ultimately being responsible for the business results of all programs for a business unit or enterprise. The specifics of a PMO leader's responsibility will depend upon how senior management charters the role.

The role has many elements of a functional manager responsible for driving excellence in the program management discipline and is focused on consistency of program management performance within the organization. Key responsibilities of the PMO leader are shown in Table 13.1.

A strong PMO leader is a person who has successfully led individuals in achieving consistent performance results, and possesses strong interpersonal skills, including the ability to interact and communicate successfully with personnel and management at all levels within the organization.

Table 13.1   PMO leader responsibilities.

PMO Leader Responsibilities	
Be directly involved in the hiring and performance reviews of program managers within the organization	Assist the program managers, when necessary, in achievement of their program objectives
Provide direction, coaching, and support to program managers and their respective teams	Provide training and development as well as advice on career development planning for the organization's program managers
Provide the tools and infrastructure necessary for program managers to do their jobs	Organize and lead program reviews for the program managers with senior management
Ensure program teams are adhering to all governance and other corporate policies and procedures	Make course correction and adjustments to programs which are missing planned targets
Disseminate periodic program status reports and metrics to senior management and other vested stakeholders	Create roadmap and plan for continual program management growth and maturity

Listed in Table 13.2 are some of the critical knowledge, abilities, and personal attributes that the PMO leader should possess. It is important to note that the competencies discussed under the program management competency model (Chapter 11), are considered as the basic skills needed for the PMO leader.

As it is with program managers, it is a rare individual who comes to the job of PMO leader fully qualified for the role. Normally, organizations seek to hire the best qualified individual available and then grow them into the role. It must be kept in mind that becoming totally proficient as a PMO leader is *a journey*. Given the continual changes in the business environment and program environment, and the need for new methods, processes and tools, we are all on a journey to improve our skills and capabilities.

MEASURING PMO EFFECTIVENESS

One of the primary challenges facing a PMO leader is to measure and communicate the effectiveness of their PMO. Unlike individual programs which have access to established metrics to measure performance and effectiveness, PMO-level metrics and measures are not well established.

Table 13.2   PMO leader knowledge, abilities, and personal attributes.

PMO Leader Knowledge, Abilities, and Personal Attributes

Knowledge	Working knowledge and experience with all program management methods and tools
	Thorough knowledge and ability to apply corporate governance and other required policies, procedures and standards
	General knowledge of the industry and the markets served by the organization
Abilities	Ability to apply solid business practices to corporate strategic and business requirements and relate these to program business results
	Capable of acquiring organizational respect and be recognized as a leader in the discipline of program management
	Capable of driving and providing the leadership for achieving the long-term viability of program management and the PMO entity within the firm
	Ability to negotiate and drive to resolution barriers and issues related to and impacting the program management discipline within the organization
	Ability to establish cross-industry network to continually look for best practices and tools to adopt
	Experience with market and customer assessment and validation tools for evaluating potential new opportunities
	Ability to stay abreast of changes and trends within the markets served
	Ability to build and sustain positive working relationships
	Capable of managing multiple, simultaneous responsibilities successfully
Personal attributes	Highly motivated and a self-starter
	Goal-oriented
	High degree of perseverance and tenacity
	Confident in one's ability to lead
	Courage to confront personnel as needed and appropriate at all levels of the organization in a professional manner
	Positive and upbeat
	Honors commitments
	Astute at finding ways to get things done with the ability to innovate when necessary

Additionally, measuring both business effectiveness and operational effectiveness can be a complex undertaking. In general, PMO-level measures and metrics must focus on the aggregation and consolidation of the program-level measures and metrics to provide a higher level assessment of organizational effectiveness.

Project offices and functional program management offices are primarily focused on operational effectiveness. Therefore, measures and metrics for these PMOs should generally be focused on operational factors such as adherence to processes, performance to schedule and cost, and predictability of project output delivery.

Business unit and enterprise PMOs are primarily focused on business effectiveness and therefore must be assessed on the combined business results of all programs within the organization over a specified period of time, as well as appropriate operational factors. Some firms have found that the assessment is helped by utilizing moving average metrics in order to normalize for variations based on time frame.

Table 13.3 provides a number of measures and metrics for evaluating effectiveness of business unit and enterprise PMOs.

Each PMO leader should discuss and come to agreement with their senior management team as to what the formal measures and metric should be to evaluate the effectiveness and continual improvement of their PMO. A focused effort is necessary to ensure that the metrics fit the needs of the business and avoid any overlap and redundancy with other metrics and measures used within the organization.

THE FUTURE OF THE PMO

There is evidence to believe that the future of the PMO looks bright. Although there are cases of both success and failure, much has been published regarding the value and benefit of effective PMOs. They have been well recognized in industry standards and are visible in every industry sector of work. It is important to note, however, that there is a strong correlation between how long a PMO has operated in an organization and its success rate.[8]

Literature tells us that the primary reasons for establishing PMOs tend to center around improving project and program success rates, and the implementation of standardized procedures, both *operational* reasons. Our personal experiences show that organizations are also turning to PMOs to help achieve repeatable *business* results, by establishing better

Table 13.3 PMO Effectiveness measures and metrics.

Measure	Description	Evaluation Method	Metric
Time-to-market (time-to-go live)	Program outputs released on schedule	Moving average percentage	Number of program outputs released on time, divided by total number of programs completed
Profit margin	Profit margin performance post release	Rolling four quarter moving average	Average profit margin one year after program output release
Profitability index	Profitability index performance post release	Rolling four quarter moving average	Average profitability index one year after program output release
Investment performance	Management of program investment	Quarterly moving average	Total budget spent for all programs versus planned investment spend
Cycle time	Improvement in program cycle time	Quarterly moving average	Average cycle time from solution concept to program output release
Program stability	Adherence to program strike zone target and threshold boundary conditions	Point calculation typically performed yearly	Number of programs hitting a Strike Zone "red condition" (senior management intervention required)

alignment between execution output and strategic intent, improving management of business risk and complexity, and building world-class program management competence. One senior executive we spoke with praised the role of the PMO within his organization:

> Program management is designed to successfully execute cross-organizational collaboration for achievement of business objectives, and the best organizational entity to manage the multitude of activities and ensure effective governance of program progress is the program management office.

Of course, this does not mean that there are not challenges to overcome. Perceptions regarding the value of PMO's are a significant problem. Additionally, establishing a PMO, or increasing the span and influence of a PMO, modifies the existing power structure and roles and responsibilities within an enterprise which breeds resistance to change. Finally, it is probably rare that a PMO leader does not experience the perception somewhere in the broader organization that a PMO is a natural breeding ground for bureaucracy when creating and implementing new policies and standards that the organization needs to abide by.

Despite these challenges, we clearly see that implementation of PMOs is again on the rise. The difference today is that the PMOs are trending more toward being lean, more focused, and better positioned within organizations. This is further reinforced in the PMI's 2013 report titled, "The Impact of PMOs on Strategy Implementation":

A PMO with a broader business-wide responsibility, such as an enterprise-wide PMO is closest to delivering business value because it is more likely to routinely help in aligning development outcomes to strategic priorities.[9]

Whether a PMO is staffed by a single individual or a multitude of individuals, it remains an innovative approach for providing leadership, infrastructure, and governance for managing and controlling multiple programs within an enterprise. It provides the ability to successfully manage and coordinate scarce resources, consistently track program and project progress, ensure that deliverables are completed correctly and, when planned, that senior leaders receive relevant and consistent information to enable good decision making, and that a firm achieves its desired business results—it's about the business.

ENDNOTES

1. Block, T. R., and J. D. Frame. "Today's Project Office: Gauging Attitudes," *PM Network*, 15 (2001): 50–53.
2. Martinelli, R., T. Rahschulte, and J. Waddell. *Leading Global Project Teams: The New Leadership* Challenge. Oshawa, Ontario: Multi-Media Publishing, 2010.
3. Archibald, Russell D. *Managing High Technology Programs and Projects*. Hoboken, N.J.: John Wiley & Sons, 2003, p. 152.
4. *The Standard for Program Management*, 3d ed. Newtown Square, PA: Program Management Institute, 2013.

5. Martinelli, Russ, and Jim Waddell. "Achieving Common Leadership and Infrastructure Through the Program Management Office." *Project and Profits Magazine* 4 (2004): 75–80.
6. Hanford, Michael F. "Program Management: Different from Project Management." www.ibm.com, May 14, 2004.
7. Tjahjani, Lia, Paul Dwyer, Mohsin Habib. *The Program Management Office Advantage*. New York: AMACOM, Division of AMA, 2009, p. 205.
8. Ibid., p. 224.
9. "The Impact of PMO's on Strategic Implementation", PMI's "Pulse of the Profession" In-Depth Report, Project Management Institute, Newtown Square, PA, 2013.

Appendices

Case Studies
in Program Management

Throughout the course of this book, we have explored program management from many dimensions. First, we explained what program management is and then how it can be used to help achieve the business strategies of an organization. We also addressed the need to align organizational business strategies with program operations and to illustrate how program management aids the company in delivering the winning value proposition in the form of the whole solution.

We then detailed the process of defining, planning, executing, and terminating a program and explained how program management practices, metrics, and tools are enforcers for successful program management. Next we turned attention toward the program manager in which we explored the roles, responsibilities, and core competencies needed to successfully manage a program. Finally, we showed how a transition to program management can be executed in an enterprise and how a PMO can serve as both a business function and operational support organization for the program management discipline.

In Part VI, we use comprehensive case studies to demonstrate how program management is applied in practice. In choosing the case studies, we looked at multiple program characteristics such as industry, program size and nature, and program management application. The cases cover multiple industries and aim to provide perspective and context regarding the use of program management in today's businesses. We have chosen not to reveal the true names of the organizations involved to ensure that the good, the bad, and the ugly aspects of each case remained intact. The cases are real, however, as are the characters and the stories contained within.

Appendix A

"I AM the PMO!"

It is a trend that is on the rise, Program Management Offices (PMOs) that are initiated and led by individuals who single-handedly champion and grow the program management discipline within their organization. This is the story of one such leader, Ron Jacobsen, who demonstrated that a one-person PMO can become a value-contributor to an enterprise through the combination of knowledge, drive, and determination.

A COMPETENCY LOST

The company highlighted in this story is a large, multi-national enterprise that provides advanced system integration prowess and a broad product portfolio to military, federal, and civil service customers.

During the first decade of the new millennium the company expanded its capabilities and offerings to meet the rapid growth in defense spending through a number of strategic acquisitions. However, each newly acquired company brought its own business systems, processes, and procedures that represented their local knowledge and were minimally documented. Decentralization was emphasized and inconsistencies in the management of programs emerged across the company. Some of the larger programs were experiencing above-plan costs and schedule delays, yet were still achieving consistent levels of customer satisfaction due to the superior technical expertise and customer service provided by the company.

One of the business units within the company experienced a significant clash of cultures as a result of the company's merger and acquisition activities. One of the merged corporations placed significant emphasis on processes, procedures, and training of its employees in program management, while the other company retained its strong engineering orientation. At the end of the acquisition process, the integrated company no longer placed the same emphasis on program management as a true discipline and function of the organization. Program management oversight was relegated to a voluntary council, which was largely ineffective because it was starved of both authority and budget. Council representatives were provided by the business areas as an additional duty rather than a priority, and their engagement was sporadic as dictated by the demands of their programs.

However, an abundant defense budget promoted a business environment where orders and sales grew at a rapid pace. Engineering managers and project leaders were promoted into program management roles based on their achievements as engineers, but were not trained in program management. As a result, programs were run as projects and were not meeting the business performance measures such as cash flow and profit margins. Program planning, if any, was the minimum required by contract. Risk management was loosely practiced and mainly by engineering. Estimates, when completed, were inaccurate, and financial performance was below plan.

Lines of responsibility, accountability, and authority had also become blurred. The environment encouraged customer intimacy where the program manager's role was largely outwardly facing. Program management processes were considered too administratively burdensome and an unnecessary cost. Project leaders ran the programs and engineering managers had significant influence on the direction of the programs. All of this led to conflicting roles between program managers and project leaders as well as higher levels of management in customer communication and contact. Long-term strategic planning became overly optimistic based on the varying views of the program manager and the levels of management above. Business planning was inaccurate and programs were not achieving their annual goals and objectives because program execution primarily remained technically focused. In short, the focus was on winning business in a low-margin, highly competitive industry.

After a number of years managing confusion, the leadership team recommended a renewed focus on program management and the establishment of a program management office to the senior executive of the business.

"WELCOME TO THE COMPANY"

The business leaders decided to reinstitute program management under a new model based upon the senior executive's belief that program managers' skills and behaviors must include business acumen, strategic focus, and solid leadership. He believed these skills were an integral part of the program manager's behavior that was not currently demonstrated in their business unit. The direction of any program requires a plan that provides a balanced, strategic approach between the customer's requirements, technological advances, and the needs of the business to realize its return. The program managers within the organization needed to understand why they owned a "business" within the business.

There was consensus among the leadership team that there was a need for a strong PMO manager who could implement this new model, and that this position had to be created with equal footing to the business unit directors to emphasize the importance of program management.

They also believed that the PMO had to be part of the senior management team at a level comparable to other critical functions such as finance, engineering, and operations. The reasons for this were:

- It establishes program management as a true function on par with the other key functions in the organization.
- It provides alignment between business strategy and program execution, consistency in business results across the programs, and development of program management competencies and career path.
- The PMO must have sufficient political and decision-making influence to broker tensions between functions.
- Increased emphasis on program management addresses two common business needs: 1) consistency in definition, planning, and execution of all programs, and 2) the need for a single program point-of-contact and accountability that provides improved communication, decision making, and program oversight.

The precise roles and responsibilities were loosely defined to enable recruiting of the right individual with a proven record of accomplishments and in-depth knowledge of program management processes. The person selected for this role needed to understand the program management discipline through direct experience as a program manager. The experience gained by leading programs of significant size and complexity was necessary to understand the range of programs in the company's portfolio.

A proven track record of leading highly technical, integrated programs while achieving cost, schedule, and other program objectives had to be evident. The PMO Manager role was defined as:

- Measuring overall success of programs and their ability to meet business objectives.
- Focusing on consistency of methods, tools, metrics, and practices across all programs and projects.
- Defining program metrics to ensure programs are meeting business objectives.
- Developing and administering a career ladder for growth and advancement of program managers in conjunction with human resources and the other business directors.
- Being the focal point for competency building and providing a learning environment for alignment of leadership, philosophies, and practices.
- Continuing the improvement and maturity of people, processes, and practices by capturing best practices, tools, and metrics.
- Establishing and evolving a central knowledge base for program management.

The company found who they were looking for in Ron Jacobsen. Jacobsen spent 27 years with a large corporation in the aerospace industry where he managed three international programs over a ten-year span. During this time, Jacobsen experienced the implementation of an executive PMO and evolution of the company's program management processes. Jacobsen also served on the steering committee for those processes and taught program management courses to program managers and other company personnel.

UNDERSTANDING THE LANDSCAPE

Jacobsen began his new role by understanding the current state of program management competency within his new company. The existing procedures were read, program reviews were attended, and the work of the program management council was assessed. He found that a comprehensive training curriculum was published; however, little budget was provided to actually conduct the training. Small pockets of coaching and mentoring based on self-assessed competencies were evident throughout the organization, but company-paid professional development was provided on a limited basis only, due to the minimal available budget.

Jacobsen then interviewed senior management, engineering and functional organization managers, and program managers to understand the nature of the environment. The consistent messages that came out of the interviews were:

- Program managers were not intimately familiar with the customer's requirements, budgeting cycles, or engineering change proposal process.
- Programs did not consistently identify risks and opportunities or include them in the estimate at completion.
- Subcontracting was unquestionably everyone's "hot button."
- A need existed for a more structured program management approach, a set of tools, and a management council to govern program management.
- There did not appear to be a baseline of program management skills.
- Program communication was critical but there were conflicting ideas on how to communicate effectively.
- New programs tended to be scattered in the early phase but got better with time.
- Uncertainty of the various levels of responsibility, authority, and accountability for a program manager was prevalent.

ESTABLISHING "STREET CRED"

The assessment of the priorities of what needed to be improved and where to start had to entail buy-in by everyone involved (program managers, the functional organizations, and the management chain). Since there was to be no PMO support staff, everyone had to share a common goal to improve the collective program management skills and abilities. However, few completely understood the role of a PMO even though there had been plenty of briefings and one-on-one meetings with the program managers in multiple geographic locations. The social attitude of "not invented here, won't work here" became evident quickly. A number of times comments such as, "We're not the company you came from" or "We're not an airplane" or "The processes don't fit the size and complexity of programs we have" were voiced. Establishing credibility with senior management and the organization's program managers was vital since Jacobsen and the PMO were new to the company and the business unit.

Shortly after joining the organization, Jacobsen demonstrated the power of program management by guiding the recovery of a program for which he had no product or customer knowledge related to this company. His success was based purely upon the reliance of thorough application of the program management discipline. Concurrent with this, he met with all program managers to explain the PMO role and vision and got their input into what made their jobs difficult. His review of procedures determined they were adequate to give a general idea of what a program manager was supposed to do. He attended program reviews to get more insight into the culture and how effectively program managers communicated program execution.

CREATING A TRANSFORMATION PLAN

Jacobsen began to establish a vision for transforming the organization based upon the executive manager's belief in the need for a strong program management capability. The vision would be based on several factors. First, the establishment of common program management processes would be required, and then understanding and applying those processes effectively would require training and education. Once trained, a program manager's competence could be measured in the effective application of those processes toward managing a program and delivering positive business results. Further development of the program manager could then be supported by additional training and education or an assignment to another program for career development. This iterative approach is displayed in the figure below.

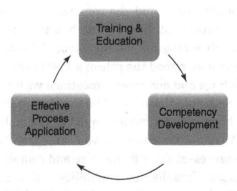

Figure A.1   A vision for building program management competency.

Jacobsen then developed and proposed a three-year transformation plan that was based on defining the job, teaching it, setting expectations for it to be followed, and measuring improvement in program execution. The primary elements of the three-year transformation plan were as follows:

Year One: Establishing the program management model

- Define the program management model and align existing procedures to it while identifying gaps or needs for new or updated procedures.
- Review and prescribe policies, procedures, tools, and metrics with functional process owners as subject matter experts paired with a program manager to ensure alignment and understanding.
- Determine application of processes, metrics, and tools appropriate to program size and category.
- Develop and publish "Knowledge Area Guide Sheets" with the above information as a starting point and continuing reference.
- Develop and conduct training to implement program management processes, tools, and metrics.
- Establish a cross-business view to optimize program manager assignments across programs through effective workforce planning.

Year Two: Positioning program management for the long term

- Overcome remaining resistance to program management acceptance.
- Continue improving program planning, execution, and control.
- Maintain excellence in customer intimacy.
- Validate program management processes and rectify gaps in or need for new ones.
- Continue the cross-business view to optimize program manager assignments across programs through effective workforce planning.
- Continue program management training in greater detail on processes and financial acumen.

Year Three: Establish program management as the business foundation

- Measure effectiveness of program planning, execution, and control.
- Grade each program's implementation of program management processes and tools.

- Adopt best practices for process improvements.
- Establish a complexity model used to determine application of processes, metrics, and tools appropriate for the level of program complexity.
- Begin filling the talent funnel with high potential junior program managers.

The transformation plan was approved by the leadership team and then rolled out to the program managers. A budget for training was estimated and approved by the executive sponsor—a sign that he was solidly behind the plan.

Defining the right program management model to be implemented at the beginning of year one would require substantial effort. In response, Jacobsen established a three-phase approach (Table A.1).

Phase I proved the most daunting considering Jacobsen did not have the domain knowledge of the range of products offered by his new company. His domain knowledge was aerospace program management and processes. He forged his expertise in the former domain by successfully managing three complex programs in an environment where processes were to be followed with discipline. It was eventually defined that there

Table A.1　Three-phase approach for year one transition.

Phase	Critical Transition Steps
Phase One	Size and define program categories.
	Define processes and align existing procedures or identify new ones.
	Review and prescribe program tools and metrics.
	Determine application of processes, metrics, and tools appropriate to program size.
Phase Two	Conduct two days of pre-requisite training to level-set and orient program managers in achieving successful business results through effective program management.
	Require training in processes, tools, and metrics and set expectations for performance through an interactive three-day workshop.
Phase Three	Finalize all elements of the transformation plan and measure implementation effectiveness.

were four common denominators to any program regardless of its size, complexity, or company:

1. It has to have a plan—plan your work, work your plan.
2. Baseline management is critical—establish them, manage to them, and control change to them.
3. Risk and opportunity management are needed—zealously drive out cost, schedule, and technical risks and exploit any and all opportunities where there is a derived benefit to the program and customer at a reasonable cost.
4. Deliver positive financial results—to the company and to the shareholders.

There were numerous challenges that defined the boundary conditions in establishing the program management model and the PMO. First, senior management decided that the PMO would not have any staff personnel supporting Jacobsen. Second, a large number of new procedures could not be generated to prevent an administratively burdensome and rigid environment that would increase program costs. Third, what was to be implemented had to fit the range of programs in the portfolio.

Jacobsen decided to use a participative approach to get buy-in across the organization and become part of the solution through the development of his plan.

GENERATING RESULTS

Two of the main criteria used for measuring the effects of the new program management model and PMO were the number of "Yellow" or "Red" programs (those experiencing critical implementation issues) within the business unit's portfolio and the value of operating margin for each program. Once the PMO was fully operational, the business unit became the only one that no longer had "Red" or "Yellow" programs. Also, a gain in its overall adjusted operating margin from 9.4 to 11.9 percent was experienced within the first three years.

Additionally, the PMO began to experience intangible results such as requests for Jacobsen's program managers to assist programs in other business units, getting consistency of results from the program managers, having a common understanding of program management processes, and the ability to move program managers between programs to adjust to the changing business conditions.

KEYS TO SUCCESS

Establishing credibility with senior management was paramount to the plan's success. Jacobsen had a supportive senior management champion whose endorsement was demonstrated early on during a "town hall" meeting in which he rolled out the PMO purpose, roles, and relationships on the part of the leadership team, and Jacobsen's three-year transformation plan. He also reiterated his endorsement frequently at business plan reviews and other meetings.

However, the sponsor could have provided further endorsement and empowerment of the PMO role by rejecting demands that the PMO prove it was adding value as an overhead function. Overhead, it was decided, was the best way to spread the cost evenly across the organization. There is a perception that PMOs are viewed as a *cost contributor* and not a direct revenue producer. But, that perception overlooks the direct leadership provided to all programs under the PMO's responsibility that are driving revenue-generating results.

Other critical factors for success identified were obtaining program manager feedback on required revisions to the procedures and processes as an indication to the program managers they were being listened to and were indeed part of the solution. Revising the monthly program review to a standardized format and instructing the program managers on what the data should tell them was a critical milestone. The monthly cost, schedule, and technical status provided indication that the quality of the program manager's knowledge was improving as was the implementation of program management processes across the board.

Ultimately, though, a large share of the credit for the improvements seen in the business unit belongs to the professional program managers who accepted the model as the change needed to enhance their skills and abilities and run their programs more effectively.

Jacobsen's three-year program management transformation plan is currently nearing completion and is being "moved upstairs." It is in the process of being expanded to the rest of the company along with Jacobsen, whose contributions have been recognized by senior management. Jacobsen now has the president of the company and the executive leadership team as sponsors.

The Office of Program Management (Jacobsen's new role and responsibility) now heads up the Program Management Council which is made up of other business units' PMOs and functional representatives. The Council now has the "teeth" to implement process and procedural changes across the enterprise as well as assess and coordinate changes in subperforming programs to improve performance. Jacobsen spends most of his time with mid-level managers explaining and reinforcing the PMO role and the resources that can be brought to bear through independent assessment and intervention on any program.

The three accomplishments that Jacobsen values most are turning a skeptical group of former engineers into professional program managers, demonstrating the effectiveness of program management processes regardless of size and complexity, and achieving business results with them!

Appendix B

LorryMer Information Technology

This is a story about the information technology (IT) department within LorryMer Corporation, a leader in the specialized motor vehicle industry in North America. LorryMer specializes in designing, developing, and manufacturing a complete line of technologically advanced motor vehicles. The company has been in business for more than six decades and now operates seven major vehicle manufacturing plants, with one plant in North America. With more than 14,000 employees, its mission is to provide the highest standard of technological innovation and premium quality to its customers.

Saul McBarney, LorryMer's IT program management officer, has been with the company for 25 years. McBarney is proud of the company's history, strategy, and background. "We are unique in terms of listening to the customers. We find out what customers' business needs are first and then develop products for them that meet their needs."

However, 2009 was a painful year for LorryMer. As a result of economic hardship in the motor vehicle industry, the company's sales and profits dropped, and its losses exceeded $1 billion. In September 2010, the *Refrigerated Transporter* reported that used automotive prices caused many transportation carriers to extend trade-in cycles because the resale prices on vehicles were less than what was owed.

As a response to the crisis, the company embarked on a major cost-saving initiative in late 2009, which later produced a number of cost-saving programs. James Ostar, a LorryMer business value account manager, said, "It was about cost, cost, cost, and nothing else."

Nevertheless, LorryMer continued to experience substantial business challenges in the following years. With the economy still suffering, LorryMer's automobile production in 2010 did not meet 2009 production levels, which were significantly lower than 2008. Funding for any new product development programs was minimal, especially during 2011. LorryMer needed to change to survive. This renewed focus on operational cost reduction elevated the strategic importance of LorryMer's IT organization. A number of significant cost reduction programs were proposed and funded by the company.

OVERCOMING STRATEGIC OBSTACLES

Bill Mennon, chief information officer for LorryMer, said, "The business environment for all automobile manufacturers remained depressed and was not expected to regain its historically high levels until late 2012. We knew we had to make some changes to be ready to compete." LorryMer focused on reducing costs quickly. As a result of relentless cost cutting programs, LorryMer's operating costs were dramatically reduced and breakeven profit was achieved in 2011.

"However, we knew those changes were tactical and not sufficient," Mennon said. "We had to make more strategic changes." Therefore, during 2013, LorryMer modified its strategy by focusing on five new core values, as follows: providing market leadership and brand coverage; pursuing technological innovation; partnering with operators for maximum productivity; focusing on the needs of its customers, employees, communities, and the environment; and being an advocate for their industry.

ALIGNING IT PROGRAMS TO BUSINESS STRATEGY

Historically, the IT strategy at LorryMer has been to maintain legacy systems. Most of the new IT investments were dedicated to e-business websites for after-market sales and marketing. Mennon said, "Eighty-five percent of what we sell is a complete commodity available from thousands of other people. So the only thing we really have to sell to increase profit is service." The remaining incremental IT spending supported enhancements to keep the legacy systems compliant with government regulations and provided some basic level of additional functions.

To improve the company's competitive position, radical changes were needed to create an IT environment that was better positioned to support all five of the new core values. As part of the improvement efforts, the IT department was reorganized and made much more agile in 2013. That was done by replacing the functional structure with a matrix organization consisting of application (IT application developers and service providers) and program management (program managers and business system analysts) competency centers. Then, a business advisory group concept was introduced. It consisted of value account managers who provided IT focus in business units and acted as the IT voice of the customer, which was something LorryMer lacked previously. Also, a joint strategic planning session between the business units and IT was initiated in the 2013. Its purpose was to determine the cost reduction needs of the business units. McBarney commented, "This was previously unheard of. In the old system, IT goals were either left to IT or imposed on us by the business units. Now, our strategic goals and direction were determined by the end users in the joint strategic planning."

To ensure IT strategy and business strategy alignment, IT activities and programs had to complement the needs of the business. The IT team used a strategy alignment chart to accomplish this (Figure B.1). Mennon explained, "The alignment chart provides a strategic mapping of the business goals, the business values (initiatives), and the IT programs. As part of the strategic changes, we were tasked by the company to design the alignment process, part of which was accomplished by the alignment chart which helped us visualize the alignment among business goals, business value, and the programs."

Mennon offered an example of how to read the chart. "A bubble on the intersection of business goal 4 and business benefit 3 means that they are aligned. Further, a bubble indicates that business value 3 intersects with program 3, meaning they are aligned. In summary, program 3 delivers business benefit 3, which achieves business goal 4. The alignment is all about IT programs contributing to business benefits by helping us achieve our business goals. That is the language we want everyone to speak."

The alignment chart is a useful mechanism that not only helps businesses visualize the alignment efforts, but also builds a standard language in the company in which every IT investment is aligned to an organizational need.

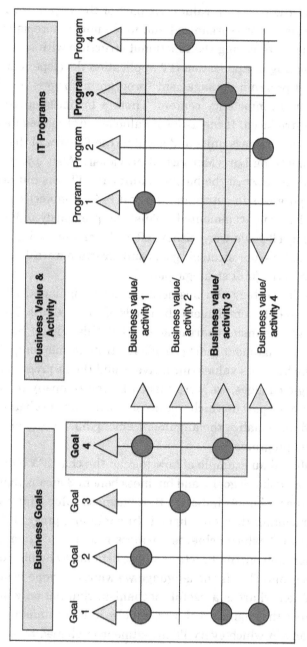

Figure B.1 Aligning IT programs to business strategy.

ALIGNMENT BEGINS AT THE TOP

On paper, the alignment chart is straightforward, but the execution of the steps in the alignment process is more complex. LorryMer has a mix of formal and informal processes for ensuring proper IT strategy and business strategy alignment. The processes are internally embedded within the business strategy formulation and throughout the life of the IT programs.

The six steps in the alignment process are as follows:

1. Strategic planning
2. Informal portfolio management
3. Program envisioning
4. Program planning
5. Program execution
6. Capability deployment

In step 1, strategic planning, the strategic plan document is developed to help accomplish the goals of IT in support of the company's business strategies and goals for the three-year-planning horizon.

According to Mennon, "Our challenge within IT has always been to take a look at how we treat the goals, whether they're well defined or not, and determine what our strategic plan is to help accomplish those goals. Our approach is an evolutionary process. Once you know where you want to go, you take it one step at a time and one program at a time, and each program gets you closer to the goals."

Then, business value, called programs, are formulated based on the goals of the business strategy. These are done by the business advisory group and Program Management Competency Center. Tools at the strategic level, which are used to ensure the quality of the alignment, include the strategic plan document, road map charts, and alignment charts, as follows:

The **strategic plan document** is a tentative capital plan that shows the snapshot of all program summaries, program types, estimated head count and payback. It is based on the business strategy and goals of different business units that IT supports.

Road map charts are used to define where the company wants to be in IT in the next three-year time frame. They address all of the business goals of different business units that IT attempts to support and their time frames.

Alignment charts are the mapping of the business goals, the business benefits and the IT programs.

Step 2, the informal portfolio management process, is a significant part of LorryMer's alignment process. Once the portfolio management process is completed, it produces information about a set of candidate programs, their investment opportunities, and program priorities. In short, the process reveals the most viable programs and possible risks that could occur in the envisioning, planning, development, and deployment phases of a program. According to McBarney, "Portfolios help to assess and monitor programs that are the best investment." Currently, this assessment is done in the portfolio process through business value assessment (BVA). BVA consists of a set of criteria prepared by value account managers. The program management office (PMO) is asked to rank the criteria regarding business-value weight, which is expressed on a 1–10 scale, with 1 being the lowest value and 10 being the highest value. Second, it assesses business risk, which is also expressed on a 1–10 scale with 1 being the highest risk and 10 being the lowest risk. An example of BVA criteria is shown in Table B.1.

Programs are then evaluated and selected based on BVA, program type, program size, priority, and business area. Once programs are selected, they go through the standard life cycle phases of envisioning, planning, development, and deployment. A program manager is usually assigned at the end of the portfolio process or the beginning of the envisioning phase, which is step 3 of the alignment process.

McBarney commented, "Ideally, we like to have a program manager on board when envisioning is just ready to begin. That allows enough time to develop a close relationship with the business system analyst. Sometimes we engage a program manager immediately at the time the portfolio process starts. Therefore, the program manager is responsible, along with the value account manager, for the business value of the program and is solely responsible for achieving the program's other requirements."

"In the envisioning and planning stages (steps 3 and 4), the ball is squarely in the program manager's court," Jeff Barrison, a program manager, said.

After the program plan is approved, the program team starts to implement step 5, or the development stage, of the alignment process and, later, the capability deployment stage (step 6). According to Barrison, program metrics are used as a mechanism to track progress to plan. At the end of each program life cycle phase, a program manager is required to present

Table B.1   LorryMer IT portfolio business value assessment.

Business-Value Weight	1-Lowest 10-Highest	Business-Risk Weight	1-Highest 10-Lowest
Generates at least $1 million in revenue per year		Necessary cost of doing business; key business processes will cease if not done	
Cost savings of at least $500,000 per year		Required to support a new product	
Improves employee communication, culture, or morale, resulting in a 10 % increase in the next annual employee survey results		Replaces legacy systems that inhibit required business process changes	
Product quality that directly affects JD Powers measurement		Requires constrained resources	
Improves dealer/customer experience, resulting in an increase of ½ percent market share		Affects existing product build capability	
Cost avoidance of at least $500,000 per year		Affects existing motor vehicle and parts sales capability	
Payback in less than 18 months		Current customer satisfaction level will decrease as measured by JD Powers	
Regulatory requirement		Competitive threat will result in the loss of market share and/or net profit	
TOTAL*		**TOTAL***	

*Note: The higher the total the better, because it represents the highest business value at the lowest business risk.

the program status to the PMO and get customer sign off to be able to proceed from one phase to the next. Therefore, customer sign off and PMO involvement are the other governance mechanisms to make sure that the program is still in line with the expectation throughout the program life cycle.

LEARNING THROUGH PRACTICE

Aligning IT programs to business strategy is new to LorryMer. Mennon remarked, "We are learning the alignment process, and I am happy with its results. For the first time, we are aligning IT programs with the business strategy—and it works."

The LorryMer experience offers a good example of the use of an integrated management system in practice to achieve strategic business objectives. Key elements of the LorryMer experience include the following:

- Alignment starts at the top—strategy drives intended business results and competitive advantage.
- To execute strategy, a company needs to consistently select, fund, and resource the programs that best align with the strategic goals and contribute the highest business value.
- Effective program management practices deliver the intended strategic business results and create business value for a company.

Appendix C

Bitten by a Rattlesnake

The Rattlesnake program was the second product development program Ken Jensen managed since coming to his current company, a leader in the computer industry. When he decided to change companies three years ago, he felt that the ten years he spent managing programs in the aerospace industry gave him a good understanding of program management and how to manage programs to success.

The first program he managed at his new company *was* a success, but proved to be much more of a challenge than Jansen had anticipated. Part of the challenge was Jensen's learning curve of the differences between product development in the computer industry versus the aerospace industry. The most notable differences were the faster pace of work and decision making in his current company and people's willingness to take considerable risk to achieve competitive advantage.

Fortunately for Jensen, he was given an excellent program team for his first program that had a strong drive for success. Jensen was elated when the majority of the team was assigned to his second program, code named "Rattlesnake."

The Rattlesnake program was projected to take 15 months with a development investment of $10 million. The program team would ramp from 5 people in the definition stage to over 50 during full development, test, and launch. The Rattlesnake product, a multi-microprocessor data center server, was considered to be highly competitive, with 10 technology advancements that were new to the industry and certain to provide product differentiation and therefore, advantage in the market.

Jensen had high regard for the senior sponsor for the program, Vivek Rondi, the business unit general manager. Rondi was an exceptional visionary, was adept at taking the appropriate amount of risk to achieve the company's goals, was a strong and effective decision maker, and fully empowered his program managers to lead their program teams. All of this positioned Jensen to succeed with Rattlesnake.

DEFINING PROGRAM SUCCESS

Through the earliest stages of the program, Jensen led the definition and planning team through a series of activities aimed at establishing a trajectory for program and business success.

First, Jensen and the Rattlesnake team established a program strike zone, which documented the business success criteria and associated measures and metrics for each criterion. A portion of the Rattlesnake strike zone is shown in Figure C.1.

Multiple conversations with Rondi were required to agree on the right measures for each of the business success factors, as well at the target and threshold values. According to Jensen, "Establishing the strike zone is probably one of the hardest parts of managing the earliest part of a program, but probably one of the most important. It definitely tested my negotiation skills!"

From his aerospace experiences, Jensen knew that establishing a strike zone for a program was necessary, but also understanding the level of risk associated with a program is crucial for several reasons. First, it is a focusing mechanism and provides guidance as to where critical program

Rattlesnake Program Strike Zone			
Critical Business Success Factors	Strike Zone		Status
Increased market share:	Target	Threshold	
• Order growth within 6 months of introduction	5%	1%	\<Green\>
• Market share increase one year after introduction	10%	5%	\<Green\>
Time to Market:			
• Program Initiation Approval	1/3/2011	1/15/2011	\<Green\>
• Product Proposal Approval	6/1/2011	6/30/2011	\<Green\>
• Engineering Release Approval	4/1/2012	4/15/2012	\<Green\>
• Product Release to Customers	5/30/2012	6/15/2012	\<Green\>
Financials:			
• Program Budget	100% of plan	105% of plan	\<Green\>
• Product Cost	$8500	$8900	\<Green\>
• Profitability Index	2.0	1.8	\<Green\>

Figure C.1 Partial Rattlesnake strike zone.

resources are needed. Second, it is used to determine the amount of schedule and budget buffers to include in the program plan. Finally, good risk management practices enable informed risk-based decision-making.

A three-tier risk assessment scale (high, medium, low) was used on the Rattlesnake program, with both probability of occurrence and severity of impact vectors evaluated. A forth category, *showstopper*, was used to identify risk events that would have a catastrophic impact to the program if they occurred, and if no risk avoidance plans were in place.

During program definition, nearly two dozen risk events were identified prior to the business case approval meeting. However, common practice within Jensen's company was to present the "Top 10" risks as part of a program business case. For the Rattlesnake program, seven risks were identified as HIGH and three were identified as MEDIUM risk.

During the course of the planning stage, a total of seventy-three risk events were identified by the Rattlesnake core team. To Jensen, this was a large number of risk events to comprehend and manage as compared to the programs he previously managed in the aerospace industry. As Jensen described it, "I was completely overwhelmed by the number of risks we were dealing with."

Following the same practice of presenting the Top 10 risk items for the program at major decision checkpoint meetings, Jensen set about analyzing and prioritizing each of the seventy-three risk items with the help of his team. During this process, many unknowns were flushed out and driven into the program plan, while the remaining unknowns were characterized and response plans were developed as appropriate. This time, the "Top 10" risks that were presented as part of the integrated program plan included two showstopper risks, six HIGH risks, and two MEDIUM risks.

Despite the existence of two showstopper risks, Rondi and the product planning committee granted approval for the Rattlesnake program to go into full development. "The approval decision really surprised me," stated Jensen. "I would never have gained approval with two potential showstoppers when managing programs in the aerospace industry," which highlighted a difference in industry risk tolerance.

HEARING THE RATTLER'S WARNING

Several months into development, Diana Best, a Rattlesnake project manager, came into Jensen's office with a concerned look on her face.

"I think we may have a problem with Tojamma Corp, the data controller device vendor," she said.

Caught a bit off guard by the comment, Jensen responded cautiously. "How so?"

Best explained that the vendor was indicating that there were some problems with the final prototype part, and that they were starting to talk about having to do a partial redesign. Jensen's caution changed to concern.

"When will we know for sure if they have to do the redesign?"

"Probably within two or three weeks," responded Best.

Trying to get a handle on the severity of the problem, Jensen continued his questioning. "And do you have any idea of the potential schedule impact?"

Best knew this question was coming, and she wished she had a firm answer for her program manager. "Unknown right now," she responded. "It depends on the extent of the redesign. I'll work to get that information, but I wanted to give you a heads-up."

"The next Rattlesnake program review with senior management is next week. Since this is one of our showstopper risks, we're going to have to be able to discuss it in detail at the review."

"Understood," replied Best. "I'll get as much information as possible for the review."

The Rattlesnake program was now five and a half months into the development stage and more than halfway through the program cycle overall. Jensen briefed Rondi on the changing status of the data controller vendor shortly after his conversation with Best. Rondi and Jensen recognized that the trigger event for the risk item had occurred: *vendor indicates a redesign may be needed*. Rondi asked Jensen to pull the engineering architect and marketing representatives together to evaluate options for the Rattlesnake product in case the data controller device is not available when needed.

As anticipated, the next Rattlesnake program review was a lively discussion centered on the highest risk item for the team: *data controller device vendor nonperformance*. Best learned that if a partial redesign of the data controller device is needed, a two to three month delay in part availability would ensue.

Rondi knew that a three-month delay in the data controller schedule will push this product outside the market window and that their competitors were not that far behind them in their own development.

Looking at Jensen, who was now at the front of the conference room, Rondi began his questioning. "What did the team come up with for redesign options, Ken?"

Having spent the morning preparing for this question, Jensen was quick to respond. "Any redesign of the existing Rattlesnake product will require two prototype builds and test cycles. This won't buy us any time." With no comments from the room, he took a quick breath and continued. "It looks like our best option is to move some of the key features of the Rattlesnake product into the existing Santiam product, which we released last year and is still doing well in the market. The team thinks this can be accomplished in one design and test cycle."

"Patricia, what's the impact to the gross margin and return on investment estimates?" asked Rondi, as he turned his attention to the program financial analyst.

"The gross margin will go down about 3 to 5 percent because we won't have the full Rattlesnake feature set to justify the premium selling price. However, we have been able to negotiate favorable prices with our current suppliers if we extend the life of their contracts by utilizing the Santiam product."

Chiming in next was the product manager, Raj Jains. "Our lead customers aren't happy, of course, but they haven't indicated they will lower or cancel their preorders if we deliver a redesigned Santiam product."

Rondi was impressed with the level of research the team did in formulating their recommendation. After briefly contemplating the recommendation of the team, Rondi provided the guidance on the next steps the team should follow. "Okay, we need to delay the start of the next program in the pipeline, and put the design resources on the redesign of the Santiam product."

"Does this mean we kill the Rattlesnake program?" queried Jensen.

"No, I want parallel efforts to continue, at least until its clear which direction we need to go," replied Rondi. "Ken, I want your vendor project manager to act like human glue to the data controller vendor. I don't want any delays in making our decision."

"Understood," said Jensen as he thought about how effective Rondi was in making the tough decisions. He still remembered how long it took to arrive at similar decisions at his previous employer.

Over the next month, work progressed on both the Rattlesnake program and the redesign of the Santiam product. Jensen and Best were in contact with each other on almost a daily basis. Best was now working in the vendor's facility in order to stay close to the internal communication. This turned out to be a wise decision.

While grabbing a cup of coffee at the vendor's office one day, Best noticed Tom Shaw sitting alone in the cafeteria. Shaw was the engineering manager for the data controller device. She decided to take the opportunity to have a quick chat about the controller.

"Hey, Tom," began Best. "How are you?"

Shaw was a little startled by Best's question, as he was deep in thought. "It's probably not a good day to answer that question."

Sensing something was bothering Shaw, she pried a little harder. "Why, what's up?"

Hesitating, Shaw decided he would tell Best the current status of the data controller. He knew she'd find out about it soon enough anyway. "I'm going to recommend a major redesign of the data controller to my management team this afternoon."

"How extensive is the redesign?" Best was trying hard to hide her anxiety.

"Nearly 30 percent of the device has to be redesigned."

"Damn," responded Best. "Yeah, I know. Please don't communicate this to your team yet. It needs to go to my senior management team first."

"All right," agreed Best. "But by the end of the day I need to be able to communicate this to my program manager."

"That's fair. I'll contact you after my meeting with the management team, if there's enough of me left, that is."

"Good luck," said Best, trying to provide some support to Shaw who was obviously dreading the events of the day ahead of him.

THE RATTLER STRIKES

The phone rang in Jensen's office at 4:30 pm. He was hoping the call was from Best because he had received some encouraging news about the redesign of the Santiam product that he wanted to share with her. "Hello, this is Ken."

"Ken, this is Diana, I have some bad news for you concerning the data controller device."

Jensen's mood changed immediately with this comment. "I have some news for you too, but it sounds like you need to go first."

Pausing briefly to collect her thoughts, Best began to tell Jensen about her conversations with the data controller engineering manager.

"The engineering manager for the data controller just left my office. He recommended a major redesign of the device to his senior management team."

"I see, and the schedule impact," asked Jensen.

"They haven't announced anything yet, but he's estimating three months," replied Best.

"Wow, that's on the high end of our estimate," said Jensen. "When are they going to officially announce the delay?"

"Probably within two weeks. They need some time to hammer out the details and draft the communications to their customers," explained Best.

"I guess my news is even more significant now," said Jensen. "The team worked out the design changes that are needed to the existing Santiam product. And they are confident that they can include all the key Rattlesnake features customers are looking for. We just don't know the time line yet."

"That's great news, Ken. Commissioning a team to go off and look at design options was a brilliant move," said Best.

"Yeah, I guess that's why Vivek is running the business; he saw this coming," replied Jensen.

"Let me know how you want me to proceed on my end, Ken."

"Will do," replied Jensen. "I'm going to try to catch Vivek before I leave today to give him a heads-up on both developments."

Jensen was hoping to talk with Rondi before leaving work for the evening as he really didn't want to sit on the news overnight. As he walked down the office corridor, he could see that Rondi's office door was still open, a sign that he was at least still somewhere in the building. As he got to the office, he saw that Rondi was on the phone. As Jensen turned to leave, he noticed Rondi motioning for him to come into his office. Within a few minutes Rondi finished his call and turned his attention to Jensen.

"Hi, Ken. What can I do for you?" asked Rondi.

"Hi, Vivek. We've had a couple of significant developments that I wanted to brief you on before you left today," replied Jensen. He then explained that he had just finished a conversation with Best, and that the data controller device had to go through a major redesign. "They haven't officially announced it yet, but the engineering manager gave her a download after he briefed his senior management team." Jensen took a deep breath, and then continued. "The impact is on the high side of what we were

Rattlesnake Program Strike Zone			
Critical Business Success Factors	Strike Zone		Status
Increased market share:	Target	Threshold	
• Order growth within 6 months of introduction	5%	1%	<RED>
• Market share increase one year after introduction	10%	5%	<YELLOW>
Time to Market:			
• Program Initiation Approval	1/3/2011	1/15/2011	<Green>
• Product Proposal Approval	6/1/2011	6/30/2011	<Green>
• Engineering Release Approval	4/1/2012	4/15/2012	<YELLOW>
• Product Release to Customers	5/30/2012	6/15/2012	<RED>
Financials:			
• Program Budget	100% of plan	105% of plan	<YELLOW>
• Product Cost	$8500	$8900	<Green>
• Profitability Index	2.0	1.8	<RED>

Figure C.2   Updated strike zone status with three month delay.

estimating—probably a three-month schedule slip, I have updated the program strike zone to see what this would mean to the program business case." Jensen and Rondi studied the updated strike zone (Figure C.2).

It was clear from the strike zone that the program business case would no longer be viable with a three-month delay to the program. Rondi collected his thoughts, and then responded. "All right, how's the team that's looking at a refresh design of the existing product coming?"

"That's the other development that I wanted to talk to you about," said Jensen. "They were able to rework the design to include many of the key features of Rattlesnake, but we don't know how long it will take yet."

"Okay," responded Rondi, looking a bit more relieved. "Go work out the most aggressive schedule possible for Santiam, assuming availability of all resources needed. Get back to me within the next two days."

"Will do," said Jensen. "What should I tell the Rattlesnake team?"

"Nothing yet. Until Tojamma officially announces the delay of the data controller device, we need to keep everyone focused on what they're presently doing." Jensen nodded in agreement.

As Jensen began his commute home, he knew he had some time to assess the current situation and think about his next steps. It was now very likely that the Rattlesnake program would be cancelled, but only if the Santiam refresh would be available within 60 days of the original Rattlesnake launch date. He knew that the company wouldn't have the market leader position with this product now, but he believed they could maintain much of the sales goals if they launched the product at the same time as their competitors.

APPLYING TRIAGE

The Rattlesnake program was entering a very awkward state between now and when Tojamma officially announced the delay of the data controller device. Jensen had a real balancing act to perform. He needed a few team members to begin planning their activities for a redesign of the Santiam product, he needed to delay work of teams that he knew would be *throw away* work if performed, and he needed to keep the remaining project teams focused on business as usual. In the next program core team meeting, he informed the program leaders that they knew Tojamma was going to announce a delay in the data controller device.

Jensen began the meeting with his announcement. "As all of you know, Tojamma has been faced with a potential redesign of the data controller device. We learned yesterday that they will indeed have to redesign a portion of the device."

Jensen paused while the program core team had a chance to absorb the news. Groans of frustration and several disparaging comments about Tojamma were expressed. The software project manager, Lynda Gould, was first to ask the question that was on everyone's mind. "How long is the delay?"

"We think as much as three months, but Tojamma hasn't officially announced their plans yet."

"Three months will kill this program," said Mike Acosta, the visibly upset validation project manager. "Are you telling us the Rattlesnake program has been cancelled?"

"No. No decisions have been made yet. Not until we get the official announcement from Tojamma. I'm sure they're trying everything they can to pull the delay in—they're aware of the impact of a three-month slip."

Jensen could see his team members relax a bit. Cancellation of the Rattlesnake program would be hard for the team to swallow. Many of them had over a year's work invested in the outcome of the program. He continued by explaining that Rondi had authorized a contingency plan with a small team to look at what it would take to put the key features from the Rattlesnake product into a redesign of the existing Santiam product and that initial assessment looked promising. Jensen informed the team that there would be a decision on the future of the Rattlesnake program within two weeks.

The decision didn't take two weeks, but two days. Tojamma announced that the redesign of the data controller device would take between eleven and fourteen weeks. Once this information was relayed to Rondi, he summoned Jensen into his office for a conference.

"Tojamma's schedule slip puts us outside of the market window," began Rondi. "The Rattlesnake program is dead in its tracks. We either get a redesigned Santiam product out there within two months of our original launch date, or we don't put a product out until next year's product cycle."

"The team believes they can get the redesigned Santiam product out sooner than two months," responded Jensen. "More like five to six weeks from the original Rattlesnake launch target."

"All right, we pull the plug on Rattlesnake and go for the redesigned Santiam product. You and I will announce it to the team in the morning. Please set up an all-hands meeting for 9:00."

"Will do." Jensen's head was still reeling from how fast the decision was made. No extended pondering, no overanalysis of "what-if" scenarios. Boom, it was done. Jensen went back to his office to send out an email about the next day's all-hands meeting.

THERE IS ALWAYS A LESSON

The basis for terminating the Rattlesnake program was because the program business case fell apart due to the failure of the development partner to meet the schedule. However, this was a documented and understood risk that was factored into executive decision making from the beginning of the program. Given this, one can debate whether this program was a success or a failure.

Jensen took away two key lessons from this program. First, it is critical to understand your program risks. The risks need to be analyzed, then prioritized based upon their potential impact on the critical business success factors and which of these risks are probable showstoppers. A good guidance system is needed to ensure a program is positioned for success and stays on course. The program strike zone was the guidance system that enabled the decision to terminate the Rattlesnake program and focus the team's efforts on delivering the redesigned Santiam product.

FINAL THOUGHTS ON PROGRAM MANAGEMENT

It's about the business! Whether your business is to make a profit for your shareholders or to provide benefit for your benefactors, this is the central program management theme carried throughout this book. When applied to its full capacity, program management provides the much-needed link between business strategy and the operational outcomes of project teams.

We began our personal journeys in program management with an understanding of how this link is established and maintained within the companies and industries in which we worked. As our journey has continued through writing, training, coaching, and consulting we have been fortunate to witness people applying the principles of program management contained within this book in ways and within industries we never imagined.

Consider some examples: the software product development company that is using program management to ensure its agile software development processes produce outcomes that remain aligned to its strategic business goals while providing iterative features to its customers; or the nonprofit energy efficiency company that is using program management to structure and manage its highly complex initiatives to achieve market transformation goals; or the product strategy and design consultancy firm that uses program management to provide stability and business focus in a highly creative design environment; or the university that uses program management to ensure curriculum is aligned to market need for the highest relevant and practical value from a student and hiring company perspective as well as meet the needs of accrediting agencies.

The utilization of program management to drive the realization of business benefits requires all actors within an organization to understand and perform their role appropriately. Whether you are a business executive, part of the middle management team, a program manager, a project manager, or an individual contributor on a program team, you have a critical role to play in order to ensure both program and business success.

We hope you have enjoyed this journey through the world of program management. We encourage you to contact us with your feedback, suggestions, and personal experiences in program management. Finally, you can find additional information, resources, and assistance at the Program Management Academy (www.programmanagement-academy.com).

Index

Want to connect?

Like us on Facebook
www.facebook.com/wileyprojectmanagement

Follow us on Twitter
@WileyPrjctMgmt

Go to our Website
www.wiley.com/go/projectmanagement